Springer Series in Chemical Physics

Volume 109

Series editors

A. W. Castleman Jr., University Park, USA
J. P. Toennies, Göttingen, Germany
K. Yamanouchi, Tokyo, Japan
W. Zinth, München, Germany

For further volumes:
http://www.springer.com/series/11752

The purpose of this series is to provide comprehensive up-to-date monographs in both well established disciplines and emerging research areas within the broad fields of chemical physics and physical chemistry. The books deal with both fundamental science and applications, and may have either a theoretical or an experimental emphasis. They are aimed primarily at researchers and graduate students in chemical physics and related fields.

Kaoru Yamanouchi · Chang Hee Nam
Philippe Martin
Editors

Progress in Ultrafast Intense Laser Science

Volume XI

Editors
Kaoru Yamanouchi
Department of Chemistry
The University of Tokyo
Tokyo
Japan

Chang Hee Nam
Department of Physics and Photon
    Science
GIST
Gwangju
Korea, Republic of (South Korea)

Philippe Martin
CEA/Physical Science Division
Lasers Interactions and Dynamics
    Laboratory (LIDyL)
Gif-sur-Yvette
France

ISSN 0172-6218   Springer Series in Chemical Physics
ISBN 978-3-319-06730-8   ISBN 978-3-319-06731-5   (eBook)
DOI 10.1007/978-3-319-06731-5
Springer Cham Heidelberg New York Dordrecht London

Library of Congress Control Number: 2006927806

© Springer International Publishing Switzerland 2015
This work is subject to copyright. All rights are reserved by the Publisher, whether the whole or part of the material is concerned, specifically the rights of translation, reprinting, reuse of illustrations, recitation, broadcasting, reproduction on microfilms or in any other physical way, and transmission or information storage and retrieval, electronic adaptation, computer software, or by similar or dissimilar methodology now known or hereafter developed. Exempted from this legal reservation are brief excerpts in connection with reviews or scholarly analysis or material supplied specifically for the purpose of being entered and executed on a computer system, for exclusive use by the purchaser of the work. Duplication of this publication or parts thereof is permitted only under the provisions of the Copyright Law of the Publisher's location, in its current version, and permission for use must always be obtained from Springer. Permissions for use may be obtained through RightsLink at the Copyright Clearance Center. Violations are liable to prosecution under the respective Copyright Law. The use of general descriptive names, registered names, trademarks, service marks, etc. in this publication does not imply, even in the absence of a specific statement, that such names are exempt from the relevant protective laws and regulations and therefore free for general use.
While the advice and information in this book are believed to be true and accurate at the date of publication, neither the authors nor the editors nor the publisher can accept any legal responsibility for any errors or omissions that may be made. The publisher makes no warranty, express or implied, with respect to the material contained herein.

Printed on acid-free paper

Springer is part of Springer Science+Business Media (www.springer.com)

Preface

We are pleased to present the eleventh volume of Progress in Ultrafast Intense Laser Science. As the frontiers of ultrafast intense laser science rapidly expand ever outward, there continues to be a growing demand for an introduction to this interdisciplinary research field that is at once widely accessible and capable of delivering cutting-edge developments. Our series aims to respond to this call by providing a compilation of concise review-style articles written by researchers at the forefront of this research field, so that researchers with different backgrounds as well as graduate students can easily grasp the essential aspects.

As in previous volumes of PUILS, each chapter of this book begins with an introductory part, in which a clear and concise overview of the topic and its significance is given, and moves onto a description of the authors' most recent research results. All chapters are peer-reviewed. The articles of this eleventh volume cover a diverse range of the interdisciplinary research field, and the topics may be grouped into five categories: ultrafast dynamics of molecules in intense laser fields (Chaps. 1–3), pulse shaping techniques for controlling molecular processes (Chap. 4), high-order harmonics generation and attosecond photoionzation (Chaps. 5–7), femtosecond laser induced filamentation (Chaps. 8 and 9), and laser particle acceleration (Chap. 10).

From the third volume, the PUILS series has been edited in liaison with the activities of the Center for Ultrafast Intense Laser Science at the University of Tokyo, which has also been responsible for sponsoring the series and making the regular publication of its volumes possible. From the fifth volume, the Consortium on Education and Research on Advanced Laser Science, the University of Tokyo, has joined this publication activity as one of the sponsoring programs. The series, designed to stimulate interdisciplinary discussion at the forefront of ultrafast intense laser science, has also collaborated since its inception with the annual symposium series of ISUILS (http://www.isuils.jp/), sponsored by Japan Intense Light Field Science Society (JILS).

We would like to take this opportunity to thank all of the authors who have kindly contributed to the PUILS series by describing their most recent work at the frontiers of ultrafast intense laser science. We also thank the reviewers who have read the submitted manuscripts carefully. One of the co-editors (KY) thanks Ms. Chie Sakuta and Ms. Ayane Maezawa for their help with the editing processes.

Last but not least, our gratitude goes out to Dr. Claus Ascheron, Physics Editor of Springer-Verlag at Heidelberg, for his kind support.

We hope this volume will convey the excitement of ultrafast intense laser science to the readers, and stimulate interdisciplinary interactions among researchers, thus paving the way to explorations of new frontiers.

Tokyo, Japan   Kaoru Yamanouchi
Gwangju, Korea   Chang Hee Nam
Saclay, France   Philippe Martin

Contents

1 Hydrogen Migration in Intense Laser Fields: Analysis and Control in Concert 1
Nicola Reusch, Nora Schirmel and Karl-Michael Weitzel
1.1 Introduction 1
1.2 Experimental Approach 3
1.3 Results 8
    1.3.1 Control by Linear Chirp 8
    1.3.2 Control by Quadratic Chirp 11
    1.3.3 Control by Other Means of Systematic Pulse Shaping 14
    1.3.4 Control by Genetic Algorithm 15
1.4 Summary and Conclusions 20
References 20

2 Electron and Ion Coincidence Momentum Imaging of Multichannel Dissociative Ionization of Ethanol in Intense Laser Fields 23
Ryuji Itakura, Kouichi Hosaka, Atsushi Yokoyama, Tomoya Ikuta, Fumihiko Kannari and Kaoru Yamanouchi
2.1 Introduction 23
2.2 Photoelectron-Photoion Coincidence Momentum Imaging 25
2.3 Channel-Specific Photoelectron Spectra 28
    2.3.1 Electronic Energy Levels of $C_2H_5OH^+$ and Appearance Energies of Product Ions 28
    2.3.2 Near-Infrared Laser Fields 28
    2.3.3 Ultraviolet Laser Fields 32
2.4 Correlation Between a Photoelectron and a Fragment Ion 36
    2.4.1 Energy Correlation Mapping 36
    2.4.2 Translational Temperature of Fragment Ions 37
2.5 Summary 39
References 40

3 Exploring and Controlling Fragmentation of Polyatomic Molecules with Few-Cycle Laser Pulses 43

Markus Kitzler, Xinhua Xie and Andrius Baltuška

3.1   Introduction 44
3.2   Controlling Fragmentation Reactions with the Shape
      of Intense Few-Cycle Laser Pulses 45
    3.2.1   Experiment 46
    3.2.2   Dependence of Fragmentation Yield on CEP 49
    3.2.3   Discussion of the Underlying Control Mechanism ... 49
    3.2.4   Recollision Ionization from Lower-Valence
        Orbitals 51
    3.2.5   Experimental Test of the Recollision Ionization
        Mechanism 51
    3.2.6   Discussion and Outlook 52
3.3   Controlling Fragmentation Reactions by Selective Ionization
      Into Dissociative Excited States 53
    3.3.1   Experiment 54
    3.3.2   Ionization Into Binding States 54
    3.3.3   Controlling Dissociation from the Cation 56
    3.3.4   Controlling Fragmentation Reactions
        from the Dication 56
    3.3.5   Selecting the Fragmentation Pathway 58
    3.3.6   Discussion and Outlook 61
3.4   Exploring Many-Electron Ionization Dynamics
      in Polyatomic Molecules 61
    3.4.1   Experiment 62
    3.4.2   Proton Spectra: Dependence on Pulse Duration
        and Intensity 63
    3.4.3   Charge State Selected Proton Spectra
        and Reconstruction of the C–H Distance 65
    3.4.4   Dependence of Charge State on Pulse Duration 67
    3.4.5   Discussion and Outlook 67
3.5   Summary 68
References 69

4 Optimal Pulse Shaping for Ultrafast Laser Interaction with Quantum Systems 73

Hyosub Kim, Hangyeol Lee, Jongseok Lim and Jaewook Ahn

4.1   Introduction 73
4.2   Types of Pulse Shaping 74
4.3   Pulse Shaping Devices 76
    4.3.1   Spatial Light Modulator 77
    4.3.2   Acousto-Optic Programmable Dispersive Filter 78
4.4   Spectral Amplitude Blocking 78

Contents ix

		4.4.1	A Ladder-Type System	78
		4.4.2	Spectral Amplitude Blocking in a *V*-Type System	80
	4.5		Spectral Chirp Control	82
		4.5.1	Chirps in a $2 + 1$ Photon Transition	82
		4.5.2	Chirps in Two-Photon Transitions	84
		4.5.3	Optimal Pulse Shaping of a Two-Photon Transition	86
		4.5.4	Chirps in a *V*-Type System	88
	4.6		Spectral Phase Programming	89
		4.6.1	Spectral Phase Programming for a *V*-Type Transition	89
		4.6.2	Spectral Phase Programming for a Non-resonant Two-Photon Transition	91
	4.7		Conclusion	92
	References			93

5 Photo-Electron Momentum Spectra in Strong Laser-Matter Interactions 95
Armin Scrinzi

	5.1		Introduction	95
		5.1.1	Size of the Computational Problem: The Infrared Curse	97
	5.2		The t-SURFF Method	98
		5.2.1	Single-Electron Systems	98
		5.2.2	Single-Ionization into Multiple Ionic Channels	101
		5.2.3	Double Ionization Spectra	104
		5.2.4	Computational Remarks	107
	5.3		Applications	107
		5.3.1	Spectra for a Short Range Potential	107
		5.3.2	Spectra for the Hydrogen Atom	109
		5.3.3	IR Photo-Electron Spectra at Elliptical Polarization	110
		5.3.4	Two-Electron System: 2×1-Dimensional Helium	110
		5.3.5	XUV Photo-Emission from He in Full Dimensionality	113
	5.4		Conclusions and Outlook	114
	References			115

6 Laser Induced Electron Diffraction, LIED, in Circular Polarization Molecular Attosecond Photoionization, MAP 119
Kai-Jun Yuan and André D. Bandrauk

	6.1	Introduction	119
	6.2	Numerical Methods	121

6.3	Diffraction in H_2 and H_2^+		124
	6.3.1	MAPDs in H_2 and H_2^+ by XUV Pulses	125
	6.3.2	Description of LIED in MPADs	129
6.4	LIED in Asymmetric HHe^{2+}		133
	6.4.1	MPAD in Asymmetric Molecules	135
	6.4.2	Interpretation of Asymmetric LIED	137
6.5	Dependence of MATI Spectra on Laser Frequency		139
	6.5.1	MPAD in MATI with XUV Pulses	139
	6.5.2	Orientation Dependent Ionization Probability	141
6.6	Conclusions		143
References			145

7 Coherent Electron Wave Packet, CEWP, Interference in Attosecond Photoionization with Ultrashort Circularly Polarized XUV Laser Pulses ... 149

Kai-Jun Yuan and André D. Bandrauk

7.1	Introduction		150
7.2	Numerical Methods		152
7.3	Electron Interference in Attosecond Photoionization		153
7.4	Electron Interference in Multiple Pathway Ionization		158
	7.4.1	Asymmetry of MPADs	159
	7.4.2	Description of Multiple Pathway CEWP Interference	162
	7.4.3	Influence of Pulse Intensity on the Asymmetry of MPADs	163
7.5	Conclusions		165
References			172

8 Phase Evolution and THz Emission from a Femtosecond Laser Filament in Air ... 175

Peng Liu, Ruxin Li and Zhizhan Xu

8.1	Introduction		176
8.2	Generation of CEP Stabilized IR Few-Cycle Laser Pulses		177
	8.2.1	Optical Parametric Amplifier	177
	8.2.2	Pulse Compression	179
	8.2.3	Carrier Envelope Phase Stability	180
8.3	Waveform Controlled THz Emission from Air Plasma Driven by Few-Cycle Pulses		180
	8.3.1	Variation of THz Waveform in Air Plasma	181
	8.3.2	Simulation of THz Emission in Air Plasma by Few-Cycle Pulses	182
	8.3.3	Variation of CEP and Phase of Few-Cycle Pulses in Filament	184

Contents xi

8.4 Initial CEP and Its Determination Through THz Waveform
Variation 186
    8.4.1 Initial CEP. 186
    8.4.2 Determination of the Initial CEP 188
    8.4.3 Experimental Verification 189
8.5 Summary 191
References 192

**9 Interaction of Femtosecond-Laser-Induced Filament Plasma
with External Electric Field for the Application to Electric
Field Measurement** 195
Takashi Fujii, Kiyohiro Sugiyama, Alexei Zhidkov, Megumu Miki,
Eiki Hotta and Koshichi Nemoto
9.1 Introduction 196
9.2 Filamentation Induced by High-Intensity Femtosecond
Laser Pulses and Its Interaction with External
Electric Field 197
9.3 Remote Measurement of Electric Field Using
Filament Plasma 198
    9.3.1 Theory 198
    9.3.2 Experimental Setup 200
    9.3.3 Experimental Results 202
9.4 Summary 209
References 211

**10 Development of an Apparatus for Characterization
of Cluster-Gas Targets for Laser-Driven Particle
Accelerations** 215
Satoshi Jinno, Yuji Fukuda, Hironao Sakaki, Akifumi Yogo,
Masato Kanasaki, Kiminori Kondo, Anatoly Ya. Faenov,
Igor Yu. Skobelev, Tatiana A. Pikuz, Alexy S. Boldarev
and Vladimir A. Gasilov
10.1 Introduction 216
10.2 Experiments and Analysis for Characterization
of the Cluster-Gas Target 217
    10.2.1 Angular Distribution of Scattered light 217
    10.2.2 Derivation of Cluster Size Distribution 219
    10.2.3 Spatial Distribution of Clusters 219
    10.2.4 Total Gas Density Profile 220
10.3 Results 221
    10.3.1 Size Measurement of Standard Particles 221
    10.3.2 Measurement of the Gas Jet Pressure 222
    10.3.3 Characterization of H_2 (70 %) + CO_2 (30 %)
Mixed-Gas Target. 225

		10.3.4	Cluster Density for H_2 (70 %) + CO_2 (30 %)	226
		10.3.5	Total Gas Density Profile	227
		10.3.6	Cluster Mass Fraction	227
		10.3.7	Size Measurement of CO_2 Clusters in Helium (90 %) + CO_2 (10 %) Mixed-Gas Target	228
	10.4	Discussion		230
		10.4.1	Comparison with the Boldarev's Model	230
	10.5	Summary		231
	References			232
Index				235

Contributors

Jaewook Ahn Department of Physics, Korea Advanced Institute of Science and Technology (KAIST), Daejeon, Korea

Andrius Baltuška Photonics Institute, Vienna University of Technology, Vienna, Austria

André D. Bandrauk Canada Research Chair/Computational Chemistry and Molecular Photonics, Laboratoire de Chimie Théorique, Faculté des Sciences, Université de Sherbrooke, Sherbrooke, QC, Canada

Alexy S. Boldarev Keldysh Institute of Applied Mathematics, Russian Academy of Science, Moscow, Russia

Anatoly Ya. Faenov Kansai Photon Science Institute, Japan Atomic Energy Agency, Kizugawa-city, Kyoto, Japan; Joint Institute for High Temperatures, Russian Academy of Sciences, Moscow, Russia

Takashi Fujii Electric Power Engineering Research Laboratory, Central Research Institute of Electric Power Industry, Yokosuka-shi, Kanagawa, Japan; Interdisciplinary Graduate School of Science and Engineering, Tokyo Institute of Technology, Midori-ku, Yokohama-shi, Kanagawa, Japan

Yuji Fukuda Kansai Photon Science Institute, Japan Atomic Energy Agency, Kizugawa-city, Kyoto, Japan

Vladimir A. Gasilov Keldysh Institute of Applied Mathematics, Russian Academy of Science, Moscow, Russia

Kouichi Hosaka Quantum Beam Science Directorate, Kansai Photon Science Institute, Japan Atomic Energy Agency, Kyoto, Japan

Eiki Hotta Interdisciplinary Graduate School of Science and Engineering, Tokyo Institute of Technology, Midori-ku, Yokohama-shi, Kanagawa, Japan

Tomoya Ikuta Quantum Beam Science Directorate, Kansai Photon Science Institute, Japan Atomic Energy Agency, Kyoto, Japan; Department of Electronics and Electrical Engineering, Keio University, Kouhoku-ku, Yokohama, Japan

Ryuji Itakura Quantum Beam Science Directorate, Kansai Photon Science Institute, Japan Atomic Energy Agency, Kyoto, Japan

Satoshi Jinno Kansai Photon Science Institute, Japan Atomic Energy Agency, Kizugawa-city, Kyoto, Japan

Masato Kanasaki Kansai Photon Science Institute, Japan Atomic Energy Agency, Kizugawa-city, Kyoto, Japan; Graduate School of Maritime Sciences, Kobe University, Kobe, Japan

Fumihiko Kannari Department of Electronics and Electrical Engineering, Keio University, Kouhoku-ku, Yokohama, Japan

Hyosub Kim Department of Physics, Korea Advanced Institute of Science and Technology (KAIST), Daejeon, Korea

Markus Kitzler Photonics Institute, Vienna University of Technology, Vienna, Austria

Kiminori Kondo Kansai Photon Science Institute, Japan Atomic Energy Agency, Kizugawa-city, Kyoto, Japan

Hangyeol Lee Department of Physics, Korea Advanced Institute of Science and Technology (KAIST), Daejeon, Korea

Ruxin Li State Key Laboratory of High Field Laser Physics, Shanghai Institute of Optics and Fine Mechanics, Chinese Academy of Sciences, Shanghai, China

Jongseok Lim Department of Physics, Korea Advanced Institute of Science and Technology (KAIST), Daejeon, Korea

Peng Liu State Key Laboratory of High Field Laser Physics, Shanghai Institute of Optics and Fine Mechanics, Chinese Academy of Sciences, Shanghai, China

Megumu Miki Electric Power Engineering Research Laboratory, Central Research Institute of Electric Power Industry, Yokosuka-shi, Kanagawa, Japan

Koshichi Nemoto Electric Power Engineering Research Laboratory, Central Research Institute of Electric Power Industry, Yokosuka-shi, Kanagawa, Japan

Tatiana A. Pikuz Kansai Photon Science Institute, Japan Atomic Energy Agency, Kizugawa-city, Kyoto, Japan; Joint Institute for High Temperatures, Russian Academy of Sciences, Moscow, Russia

Nicola Reusch Fachbereich Chemie, Philipps-Universität Marburg, Marburg, Germany

Hironao Sakaki Kansai Photon Science Institute, Japan Atomic Energy Agency, Kizugawa-city, Kyoto, Japan

Nora Schirmel Fachbereich Chemie, Philipps-Universität Marburg, Marburg, Germany

Armin Scrinzi Ludwig Maximilians University, Munich, Germany

Igor Yu. Skobelev Joint Institute for High Temperatures, Russian Academy of Sciences, Moscow, Russia

Kiyohiro Sugiyama Interdisciplinary Graduate School of Science and Engineering, Tokyo Institute of Technology, Midori-ku, Yokohama-shi, Kanagawa, Japan

Karl-Michael Weitzel Fachbereich Chemie, Philipps-Universität Marburg, Marburg, Germany

Xinhua Xie Photonics Institute, Vienna University of Technology, Vienna, Austria

Zhizhan Xu State Key Laboratory of High Field Laser Physics, Shanghai Institute of Optics and Fine Mechanics, Chinese Academy of Sciences, Shanghai, China

Kaoru Yamanouchi Department of Chemistry, The University of Tokyo, Tokyo, Japan

Akifumi Yogo Kansai Photon Science Institute, Japan Atomic Energy Agency, Kizugawa-city, Kyoto, Japan

Atsushi Yokoyama Quantum Beam Science Directorate, Kansai Photon Science Institute, Japan Atomic Energy Agency, Kyoto, Japan

Kai-Jun Yuan Laboratoire de Chimie Théorique, Faculté des Sciences, Université de Sherbrooke, Sherbrooke, QC, Canada

Alexei Zhidkov Electric Power Engineering Research Laboratory, Central Research Institute of Electric Power Industry, Yokosuka-shi, Kanagawa, Japan

Chapter 1
Hydrogen Migration in Intense Laser Fields: Analysis and Control in Concert

Nicola Reusch, Nora Schirmel and Karl-Michael Weitzel

Abstract The dissociative ionization of ethane leading to the formation of H^+, H_3^+ and CH_3^+ ions has been studied in femtosecond laser fields. The analysis of kinetic energy distributions of the ions provides information on the mechanism of the photodynamics involved. Important ingredients of these dynamics are hydrogen migration, charge localization and charge separation. We demonstrate that it is possible to go one step beyond analysis towards the control of product yields. Ultimately, controlling the yields of branching ratios provides us with even improved understanding of the competing reaction pathways.

1.1 Introduction

In recent years we have seen major progress in the field of photodynamics of molecules and molecular ions induced by intense ultra-short laser fields [1–7]. Not only have we increased our understanding of bond-breaking, but also that of bond-making [8, 9]. A prototypical example in this context is the formation of H_3^+ ions in the dissociative ionization of several small hydrocarbon molecules employing femtosecond laser pulses [10–13]. The equilibrium geometry of the H_3^+ ion is that of an equilateral triangle [14]. Forming this species from a saturated hydrocarbon molecule obviously involves the breaking of three C–H bonds and also the formation of one two-electron-three-center bond. Let us recall the case of fs-laser ionization mass spectrometry (fs-LIMS) of ethane. Here, the dissociative ionization leads to a broad range of ion signals arising in part from single ionization, but in part also from multiple ionization. One of the fragment ions is the H_3^+ ion, whose formation, as we have shown in previous work, proceeds on the doubly charged potential energy surface [13]. The doubly charged ethane ion, $C_2H_6^{++}$, has a well bound equilibrium

---

N. Reusch · N. Schirmel · K.-M. Weitzel (✉)
Fachbereich Chemie, Philipps-Universität Marburg, 35032 Marburg, Germany
e-mail: weitzel@chemie.uni-marburg.de

K. Yamanouchi et al. (eds.), *Progress in Ultrafast Intense Laser Science XI*,
Springer Series in Chemical Physics 109, DOI: 10.1007/978-3-319-06731-5_1,
© Springer International Publishing Switzerland 2015

geometry which can be viewed as a CH_4^{++} ion attached to a CH_2 moiety [15]. Since the initial ionization is a vertical event, visiting that region of phase space, where the equilibrium geometry is located, requires a hydrogen migration. This at the same time constitutes a certain time requirement for this process to occur. The process also involves the rearrangement of electron density with aspects of charge localization and charge separation. Early hints at the relevance of hydrogen migration in fs-laser ionization came from isotope labeling experiments [16]. E.g. the dissociative ionization of deuterated methanol CH_3OD leads to the observation of both H_3^+ and H_2D^+, where the latter again requires hydrogen/deuterium migration [17, 18]. In order to better understand the photodynamics of ethane, we have employed a variety of different techniques in the past. Our first experimental investigation was focused on the average kinetic energy release (KER) in the dissociative ionization of ethane leading to the formation of H_3^+ and $C_2H_3^+$ [13]. At the relevant laser pulse conditions ($>80\,\mu J$) the kinetic energies of H_3^+ and $C_2H_3^+$ were momentum matched indicating that the two ions were formed in the Coulomb explosion of a common precursor, i.e. the $C_2H_6^{++}$ ion. This interpretation was supported by the first complete reaction path obtained by means of high level ab initio calculations leading from the ethane dication to the fragments $H_3^+ + C_2H_3^+$ [13].

Subsequently, we further confirmed this mechanism by means of an ion-ion-coincidence investigation of the correlated formation of the fragments H_3^+ and $C_2H_3^+$ [19]. In that work we also detected coincidences between other pairs of fragments, e.g. H^+ and $C_2H_5^+$. At first glance it appeared surprising that the $H^+ + C_2H_5^+$ channel exhibited a clear anisotropy, indicating fast decay of the precursor, yet the intensity was much lower than that for the $H_3^+ + C_2H_3^+$ channel, which was isotropic, hence indicating a lifetime of the precursor longer than a rotational period. This puzzle can be resolved by realizing that most of the H^+ signal in fact does not correlate with $C_2H_5^+$ but with $C_2H_3^+ + H_2$ [20]. Thus, the H^+ ions observed in coincidence with $C_2H_5^+$ constitute a minority channel, although formed with a very short time constant. The formation of H_3^+ ions on the other hand, which certainly occurs in coincidence with $C_2H_3^+$ ions, is a majority channel, which involves the migration of hydrogen atoms over several energetically low lying transition states indicated in Fig 1.1 before passing the final rate-limiting transition state leading to product formation.

In this work we complement previous investigations aimed at understanding the molecular dynamics involved in the dissociative ionization of ethane by studies aimed at controlling the yields of the different molecular processes involved. We will first review very recent work employing *linear* chirp as a tool for controlling fragmentation yields [21]. We will further complement that work by employing *quadratic* chirp as well as the concept of genetic algorithms. Experimentally we will not only present ion yields but also kinetic energy distributions (KED) as a function of the laser pulse parameters.

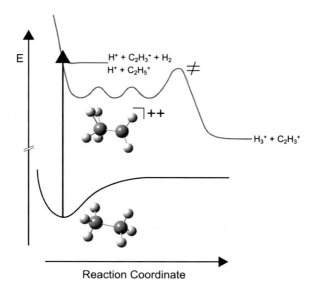

Fig. 1.1 Schematic illustration of the potential energy curves leading from the equilibrium conformation of the neutral ethane (*lower curve*) to the formation of H^+ and H_3^+ on the dicationic state. The characteristics of the potential energy curve for the H_3^+ formation illustrates the corresponding hydrogen scrambling and the transition states involved

1.2 Experimental Approach

All experiments have been carried out employing femtosecond laser pulses delivered by a multipass amplifier (Quantronix, ODIN) operated at a repetition rate of 1 kHz and seeded by a Ti:Sapphire oscillator (SYNERGY, Femtolasers). The wavelength of these pulses is typically centered at about 810 nm. The spectral bandwidth is on the order of 40 nm. The spectral phase was controlled by using a spatial light modulator (SLM-S640, Jenoptik, 640 pixel, nematic liquid crystal mask, 430–1,600 nm) in a folded 4f-shaper setup. The shortest laser pulses accessible with this laser system are typically 45 fs. The laser pulses were characterized using a frequency-resolved optical gating technique (FROG, GRENOUILLE 8-50, Swamp Optics). Experiments running the GRENOUILLE in the spatial mode confirm that a spatial chirp, which could possibly be introduced in the 4f-shaper, is negligible. The laser pulse energies used in the current study are in the range between 10 and 150 µJ, which, when focused by an $f = 75$ mm mirror (leading to a focal radius of 40 µm, as measured by the knife-edge technique), correspond to 4.4×10^{12} and 6.6×10^{13} W/cm^2 in the case of the 45 fs pulses. These laser pulses are focused into the ion source of a home-built time-of-flight mass spectrometer (ToF-MS), which mainly consists of three electrostatic lenses with a distance of 14 mm to each other and a field free drift tube (59.7 cm). The field strength in the ion source was 250 V/cm. All ions are detected by a microchannel plate device, all signals are processed in a personal computer employing the LabVIEW suit of programs. The sample (ethane, Fluka, 99.95 % purity) is introduced effusively into the experimental chamber.

The pivotal property for characterizing femtosecond laser pulses is the phase. Here, we choose to discuss the spectral phase of the laser field, which can be expanded as a Taylor series:

$$\varphi(\omega) = \varphi_0 + (\omega - \omega_0) \cdot \frac{1}{1!} \cdot \left.\frac{\partial \varphi}{\partial \omega}\right|_{\omega = \omega_0} + (\omega - \omega_0)^2 \cdot \frac{1}{2!} \cdot \left.\frac{\partial^2 \varphi}{\partial \omega^2}\right|_{\omega = \omega_0}$$

$$+ (\omega - \omega_0)^3 \cdot \frac{1}{3!} \cdot \left.\frac{\partial^3 \varphi}{\partial \omega^3}\right|_{\omega = \omega_0} + \cdots \tag{1.1}$$

The third term in (1.1) describes a quadratic spectral phase which causes a linear chirp. The linear chirp parameter α is given by

$$\alpha = \frac{1}{2!} \cdot \left.\frac{\partial^2 \varphi}{\partial \omega^2}\right|_{\omega=\omega_0} \tag{1.2}$$

By writing a quadratic spectral phase on the spatial light modulator (SLM) in the 4f-shaper setup, also shown in Fig. 1.2, a linear chirp is imprinted onto the laser pulse. This is accompanied by a temporal stretching of the pulse according to (1.3), where τ_0 is the pulse duration (full width half maximum, FWHM) of the shortest pulse achieved [22].

$$\tau^2 = \tau_0^2 + \left(\frac{8 \cdot \ln 2 \cdot \alpha}{\tau_0}\right)^2 \tag{1.3}$$

In the same manner a quadratic chirp (third order dispersion, TOD) is generated by writing a third order spectral phase onto the SLM. The quadratic chirp is described by the fourth term in the Taylor series and characterized by the chirp parameter β given by

$$\beta = \frac{1}{3!} \cdot \left.\frac{\partial^3 \varphi}{\partial \omega^3}\right|_{\omega=\omega_0} \tag{1.4}$$

Pure quadratic chirp can be considered to introduce pre-pulses, respectively post-pulses. A good overview of pulse shapes and their description is given in [23]. We will present a visualization of typical laser fields for pulse shapes used in this work in the results section.

While the *linear* and the *quadratic* chirp discussed above may be considered as systematic pulse shaping, we have also performed experiments where we searched for the optimum laser field maximizing certain ion yields. The properties of the laser fields which are varied in order to maximize a specific objective, e.g. the ion yield, are the spectral phase, i.e. the phase in the frequency domain. The experimental

1 Hydrogen Migration in Intense Laser Fields: Analysis and Control in Concert   5

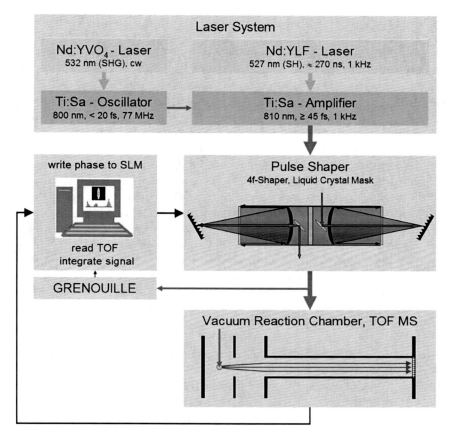

Fig. 1.2 Schematic setup of the experiment including the laser system, the pulse shaper with the spatial light modulator (SLM), the time-of-flight mass spectrometer and the data acquisition system. The main technique for pulse characterization is frequency resolved optical gating (FROG, GRENOUILLE)

device for varying the spectral phase is the SLM which is part of the 4f-shaper setup indicated in Fig. 1.2. We optically expand the laser pulse over as many pixels on the SLM as possible. Each pixel can be assigned an independent phase value between 0 and 2π. This spans an enormous parameter space, which we cannot scan manually. Instead we use a genetic algorithm to solve this optimization problem. The algorithm used by us is one delivered with the SLM but was adapted to our experimental setup [24]. The actual program used consists of two parts, one to control the data acquisition and the second one to execute the genetic algorithm.

In the context of this optimization task a spectral phase is defined by a set of 640 spectral phase parameters (genes, one per pixel). This is termed an individual in evolutionary algorithms. The first part of the program writes this spectral phase to the SLM. For each individual 500 ToF mass spectra are recorded and averaged. A baseline correction is performed and the observable of interest is extracted from this spectrum, e.g. the integrated H_3^+ ion yield. This integrated ion yield then serves as the fitness value for the genetic algorithm. The next higher level of hierarchy is the generation. Each generation typically consists of 40 individuals, for which the ToF measurement and analysis is repeated. The second part of the program analyzes these 40 fitness values and creates a new generation of spectral phases, which closes the feedback loop and restarts the first program part. The crucial part of any genetic algorithm is the protocol which decides which spectral phases will be kept for the next generation and which will be modified. This protocol will determine how fast the algorithm converges. The algorithm is considered converged once the optimum fitness parameter cannot be further increased within a range of values and loops defined by the operator. The protocol converging well on one objective may converge slowly on another. A converged result, i.e. here a spectral phase consisting of 640 entries for the 640 pixels, should be independent of the protocol chosen.

At the beginning of any optimization, neighboring parameters were binned together in five packs each containing 128 pixels. All parameters in one bin were given the same value. The binning size was decreased in the progress of the optimization between successive generations. This procedure is used to quickly approximate a rough structure in the parameter pattern and thus to speed up the convergence behavior. In general, the binning size reaches a value of 1, i.e. each pixel is addressed independently of all neighboring pixels, after 28 generations. After obtaining the fitness values from the first program part for each generation the fitness of all individuals were sorted accordingly. The best seven individuals (elites) were transferred to the next generation without further modification. Other individuals were modified or crossbred. Optimal fitness values were usually obtained after around 40 generations.

As we will describe in more detail in the result section, we observe the presence of fast and slow CH_3^+ ions in our ToF distributions. The GA is also used for maximizing the ion yield ratio of these fast versus slow CH_3^+ ions. In general distinct peak maxima are observed for fast and slow CH_3^+ ions, both in forward and backward direction. Yet, the peaks are intrinsically in general not fully resolved. For this reason integration of the ion ToF spectra is not appropriate. Instead we simply pick the peak maxima from the ToF distributions. While forward and backward scattered signals were analyzed separately for the kinetic energy (KE) analysis described below, in the case of maximizing the ratio of fast versus slow ions forward and backward ToF signals were added. Thus, the peak maxima are added for the low kinetic energy CH_3^+ ions (Y_{max}(lower KE) $=$ $Y_{max, forward}$(lower KE) $+$ $Y_{max, backward}$(lower KE)) and the high kinetic energy CH_3^+ (Y_{max}(higher KE) $=$ $Y_{max, forward}$(higher KE) $+$ $Y_{max, backward}$(higher KE)) separately. The ratio of the two quantities, either Y_{max}(lower KE)/Y_{max}(higher KE) or Y_{max}(higher KE)/Y_{max}(lower KE), then defines the fitness parameter for this particular optimization task. If Y_{max} for either the slow or the fast CH_3^+ was below a certain threshold the fitness for this individual was set zero.

As mentioned above all ions formed in the laser-matter interaction are analyzed and detected in a home-made linear time-of-flight mass spectrometer [25]. The classical ToF equations imply that a thermal kinetic energy distribution of ions will lead to a Gaussian ToF shape [26]. As a particular KE is connected to a corresponding set

1 Hydrogen Migration in Intense Laser Fields: Analysis and Control in Concert 7

of velocity vectors with ends on the surface of a sphere, a mono-energetic KER on the other hand will translate into a rectangular ToF shape [27, 28]. In the latter case, the center of this rectangular peak, observed at t_0, corresponds to ions with velocity vectors perpendicular to the ToF axis. The minimum and maximum ToF observed then correspond to ions with initial velocity vectors parallel to the ToF axis.

The width of the rectangle in the ToF domain, Δt, scales with the square root of the initial kinetic energy, $E_{kin,0}$, as given in (1.5), where m is the ion mass, q the charge, and G_1 the field strength in the ion source [$(U_1 - U_2)/d$ where U_1 and U_2 are the voltages of the first and the second electric lenses, respectively] [27].

$$\Delta t = \frac{\sqrt{2m}}{q \cdot G_1} \sqrt{E_{kin,0}} \tag{1.5}$$

This $E_{kin,0}$ can e.g. originate from the thermal properties of the sample, but also from intra-molecular Coulomb repulsion in the dissociation of a dication.

Ultimately, any ion ToF distribution can be reconstructed from a weighted sum of rectangles, where the width of the rectangle reflects the relevant kinetic energy, the height reflects the weighing factor, which e.g. also accounts for the proper ToF— energy conversion. A thermal KE distribution will thus transform to a Gaussian ToF distribution.

If the kinetic energy is large enough, ions with an initial velocity vector perpendicular to the ToF axis do not reach the detector before they travel off the spectrometer axis so far that they do not hit the sensitive area of the MCP detector and consequently are discriminated in our spectrometer setup. As described by (1.6) ions are discriminated, if the perpendicular component of the velocity $v_\perp$ is larger than the detector radius divided by the time of flight for ions with zero initial velocity parallel to the detector axis t_0. For a given absolute initial velocity ions are seen by the detector if they are emitted into a cone with specific solid angle. Angular discrimination often leads to dips in the ion ToF distribution giving rise to *forward* and *backward scattered* ions (ions with velocity vectors initially pointing towards or away from the detector). Since t_0 depends on the acceleration fields applied in the ion source, angular discrimination is more pronounced for small draw out fields than for large fields.

$$v_\perp > \frac{r_{Detector}}{t_0} \tag{1.6}$$

To deduce the energy distribution corresponding to a specific ToF distribution a genetic algorithm based on the same principles as mentioned above can be applied. Ideally, the ToF distributions are symmetric in forward and backward direction. Clearly this is not the case for the current setup (c.f. Fig. 1.11), where the effective angular discrimination is slightly different for forward and backward direction. We take account of this fact by analyzing the forward and the backward part of the ToF distributions separately and averaging the result afterwards.

The parameter set of each individual, which is to be optimized, is the probability for observing ions at a specific kinetic energy. Each kinetic energy is transformed into a rectangle in the ToF domain according to the equations discussed above. Summing up these rectangles ultimately leads to a calculated ToF distribution. The size of the parameter set depends on the maximum kinetic energy of the ions considered and the energy increments chosen. The latter is typically $\Delta E = 100$ meV for H_3^+, and 30 meV for CH_3^+.

In the first part of the program the relevant region in the ToF spectrum is selected. The edge of this range defines the maximum kinetic energy to be considered in the KED. Each KE considered is transformed into a rectangle in the ToF domain. In doing so, the angular discrimination of ions discussed above must be considered. The ToF for which the ions do not contribute to the temporal distribution are calculated and given an ion yield of zero. For all ToF for which the ions reach the detector a weighting factor reflecting the amount of ions per ToF at a given energy is taken into account.

As the goal of the optimization is a good agreement between the simulated and measured ToF signals, the fitness value is defined as the inverted root mean square (rms) between experimental ion yields and the simulated ion yields of the parameter set. Since the calculated ToF axis in general has non-constant increments (we arbitrarily chose to start with constant increments in the energy domain), we finally have to interpolate the experimental ToF data in order to be able to calculate the rms discussed above.

The fitness value obtained by this procedure is handed over to the second program part. The feedback and the generation of new parameter sets for the next generation follow the concept of the genetic algorithm method discussed above. For obtaining energy distributions, usually one generation contains 50 individuals, from which 15 are transferred to the next generation as elites without modification. Convergence is generally achieved after several ten thousand generations.

1.3 Results

This section is organized as follows. We start by discussing aspects of controlling ion yields by means of linear chirping the laser fields. This subsection will review recent data from [21]. In a second subsection we then turn to the control of ion yields employing quadratic chirp. We also briefly discuss some other approaches for systematically shaping femtosecond laser fields. Finally, in order to complement these systematic chirp variations we present results of ion yield control and kinetic energy analysis employing a genetic algorithm in a fourth subsection.

1.3.1 Control by Linear Chirp

As outlined in the introduction previous work had established a complete reaction path for the formation of H_3^+ ions from ethane elucidating the role of hydrogen migration in dissociative ionization [13, 19]. In order to shed more light on the electron and nuclear dynamics involved we decided to perform additional studies

1 Hydrogen Migration in Intense Laser Fields: Analysis and Control in Concert

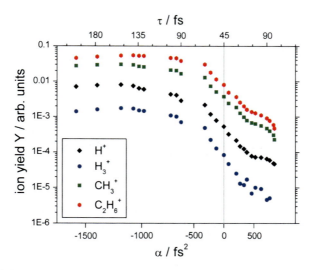

Fig. 1.3 Ion yields observed in the dissociative ionization of ethane as a function of the linear chirp parameter α (*lower axis*) and as a function of the pulse duration (*upper axis*). Note the logarithmic scale of the y axis. Data adapted from [21]

aiming at the control of branching ratios in the dissociative ionization of ethane. Here, it seems most appropriate to start by investigating the role of linear chirp in the dissociative ionization. Figure 1.3 shows the variation of ion yields with the linear chirp parameter α. Evidently the ion yields vary significantly with the sign as well as the value of the linear chirp parameter α [21]. All ion yields are maximized for values of α around $-1,200\,\text{fs}^2$.

This is a rather unique observation since most reports in the literature indicate that ion formation, in particular fragment ion formation, is intensity driven [29–31]. The latter, however, is not surprising for investigations conducted in the regime of $10^{14}\,\text{W/cm}^2$ and above. In contrast, our own studies showing pronounced chirp effects were performed at peak intensities on the order of $10^{12}\,\text{W/cm}^2$. Here, the dynamics is not in the strong field regime. On a qualitative basis we assume that a negative chirp is favorable for climbing an effectively anharmonic ladder of states compared to a positive chirp. For very high (positive and negative) chirp values the laser peak intensity decreases and leads to the observed decrease in ion yields. This is also the case for high negative chirps as this effect becomes more important. Clearly, further theoretical efforts are required in order to explain this chirp dependence. Ideally these calculations would combine high level theory for ionization as well as the dissociation of the ions.

Beyond this, also the ratio of e.g. H_3^+ to H^+ yields can be manipulated by choosing the appropriate chirp parameter [21]. The precondition for being able to control ion yields by means of chirping is again rather low laser intensity, on the order of $10^{12}\,\text{W/cm}^2$. At higher laser intensities, the dynamics involved become intensity driven implying maximum ion yields for the shortest pulse respectively the highest

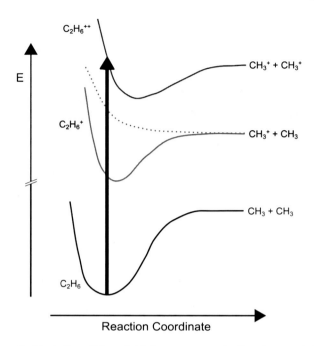

Fig. 1.4 Schematic illustration of the potential energy curves leading from neutral ethane (*lower curve*) to the formation of CH$_3^+$ ions either on the monocationic charge state (*middle curves*, note, there is certainly a bound potential energy curve but maybe also a repulsive one) or on the dicationic charge state (*upper curve*)

laser pulse peak intensities. Thus, the value of the chirp parameter α, for which the total ion yield is maximized, changes from about −1,000 fs^2 to zero when increasing the laser intensity from about 10^{12} W/cm^2 to a few times 10^{13} W/cm^2. Interestingly, the amount of fragmentation observed also increases with increasing negative chirp. For the lowest laser intensities accessible (15 μJ focused by a f = 75 mm lens) the ratio of yields Y(H$_3^+$):Y(H$^+$) is maximized for a chirp of α = −700 fs^2. Since the two ions predominantly originate from the same charge state as illustrated in Fig. 1.1, i.e. the dication state, we discussed this as an example of an *intra-charge-state* control, implying aspects of the control and localization of electron density.

On the other hand, the formation of CH$_3^+$ ions may proceed either on the same dicationic state or on the monocationic state as illustrated in Fig. 1.4. The two different charge states transform to significantly different average KER into the fragments, one being on the order of 2.7 eV, the other on the order of 0.7 eV [21]. While the KER of 2.7 eV is in line with the Coulomb explosion of a doubly charged ethane ion, the KER of 0.7 eV most likely involves impulsive fragmentation from a repulsive state of the mono charged cation. The dissociation on the ground state of the ethane monocation would lead to a thermal KER not compatible with the findings. Ultimately we were able to demonstrate that linear chirp allows to control the yields of slow CH$_3^+$ ions versus fast CH$_3^+$ ions. We consider this to be an example of *inter-charge-state* control.

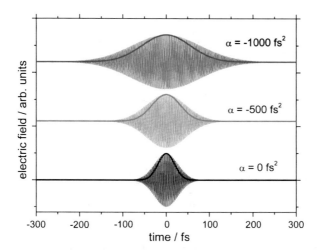

Fig. 1.5 Illustration of laser fields for an unchirped laser pulse (*lower trace*), $\alpha = -500\,\text{fs}^2$ (*middle trace*) and $\alpha = -1{,}000\,\text{fs}^2$ (*upper trace*). Center wavelength is 810 nm and the FWHM of the shortest pulse is 45 fs. Also shown is in each case the temporal intensity as a thick *solid line*. All traces normalized to the same arbitrary value, i.e. 1

To conclude this subsection on the role of linear chirp it appears helpful to visualize typical laser fields employed in the studies presented above. To this end, Fig. 1.5 shows these laser fields for the 45 fs pulse, for $\alpha = -500\,\text{fs}^2$ and for $\alpha = -1{,}000\,\text{fs}^2$. Note, the traces have all been normalized to the same arbitrary maximum amplitude of 1, in order to highlight the variation in pulse duration. In addition Fig. 1.5 also includes the three temporal intensity profiles, again normalized to 1. With increasing chirp parameter the pulse duration increases quadratically as can be seen from (1.3).

1.3.2 Control by Quadratic Chirp

As we have demonstrated in previous work [21] and also in the previous section, the dynamics of ethane ionization/dissociation can be strongly influenced by the linear chirp parameter α. Are higher order contributions to the spectral phase of similar importance? In order to answer this question we have performed a systematic variation of quadratic chirp (TOD) β for various values of α.

Figure 1.6 presents yields for the parent ion $C_2H_6^+$, as well as fragment ions CH_3^+, H_3^+ and H^+. Evidently, all ion yields shown exhibit a maximum at values $\beta = -8{,}333\,\text{fs}^3$. A minimum is observed for all ions analyzed at values $\beta = +2{,}500\,\text{fs}^3$. For positive values of β the ion yields exhibit another, local maximum at $\beta = +16{,}666\,\text{fs}^3$. Overall, the yield curves are highly asymmetric with respect to $\beta = 0\,\text{fs}^3$. Large β parameters correspond to pulse sequences effectively spread out over longer times, thus reducing the actual light intensity. In Fig. 1.7 we

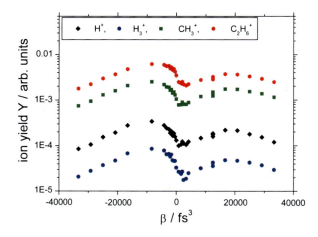

Fig. 1.6 Ion yields of some species observed in the ionization/dissociation of ethane as a function of the quadratic chirp parameter β. Note the logarithmic y axis. The laser pulse energy was 15 μJ/pulse, $\alpha = 0\,\text{fs}^2$

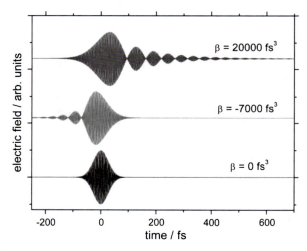

Fig. 1.7 Illustration of laser fields for an unchirped laser pulse (*lower trace*), $\beta = -7,000\,\text{fs}^3$ (*middle trace*) and $\beta = +20,000\,\text{fs}^3$ (*upper trace*). Center wavelength is 810 nm, the FWHM of the unchirped pulse is 45 fs

again illustrate laser fields for typical values of $\beta = 0\,\text{fs}^3$, $\beta = -7,000\,\text{fs}^3$, and $\beta = +20,000\,\text{fs}^3$. E.g. for $\beta = +20,000\,\text{fs}^3$ the effective laser field is spread out over several 100 fs. Note that a positive sign of β implies a distortion towards positive time (i.e. trailing part of the light-matter interaction), a negative sign of β implies a distortion towards negative time (i.e. leading part of the light-matter interaction). The decrease of ion yields for very large positive or negative β may be explained by

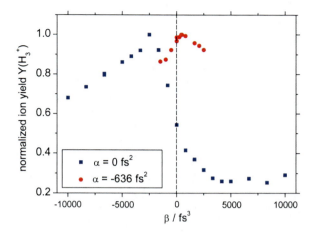

Fig. 1.8 H_3^+ ion yields observed in the ionization/dissociation of ethane as a function of the quadratic chirp parameter β for $\alpha = 0\,\text{fs}^2$ and for a negative linearly chirped laser pulse. Pulse energy: 50 μJ/pulse

the concomitant decrease in light intensity. The case $\beta = 0\,\text{fs}^3$ is not a particularly favorable spectral phase in terms of ion yields.

It appeared appropriate to start the discussion of TOD effects by setting all other contributions to the spectral phase equal to zero, i.e. looking at pure TOD effects. The obvious next question is: are TOD effects still important if we set the linear chirp parameter to a value where one of our objectives is maximized, e.g. the H_3^+ ion yield or the ratio of yields $Y(H_3^+) : Y(H^+)$. For this purpose additional experiments have been performed at slightly increased pulse energy (50 μJ/pulse). Figure 1.8 shows the H_3^+ ion yield as a function of β for a linear chirp parameter $\alpha = -636\,\text{fs}^2$, a value close to the one for which the H_3^+ ion yield was maximized for a pure linear chirp [21]. We note that for $\alpha = -636\,\text{fs}^2$ the maximum of the H_3^+ ion yield is observed very close to $\beta = 0\,\text{fs}^3$. Thus one may conclude that the linear chirp is more important than the quadratic chirp. For comparison the graph also shows the β variation for $\alpha = 0\,\text{fs}^2$. It appears that the entire data set has been shifted to more positive β values compared to Fig. 1.6.

The laser fields associated with the maxima observed in Fig. 1.8 are displayed in Fig 1.9. Evidently, the laser field for chirp parameters $\alpha = 0\,\text{fs}^2$, $\beta = -2,500\,\text{fs}^3$ is very different from that for $\alpha = -636\,\text{fs}^2$, $\beta = 500\,\text{fs}^3$. On the other hand the latter is hardly distinguishable from the laser field for parameters $\alpha = -636\,\text{fs}^2$, $\beta = 0\,\text{fs}^3$. This lends further support to the conclusion that the linear chirp parameter is more important than the quadratic chirp parameter in the dissociative fs-laser ionization of ethane.

There are rather few reports on systematic investigations of quadratic chirp in the literature, perhaps because of its subordinate role. As one example we note the work by Wollenhaupt et al. demonstrating the importance of quadratic chirp in photoionization dynamics governing the material processing relevant for microprocessing [32].

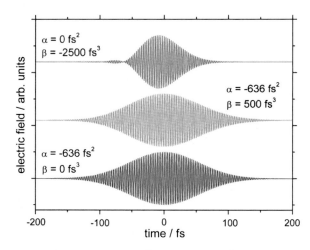

Fig. 1.9 Illustration of laser fields for $\alpha = 0\,\text{fs}^2, \beta = -2,500\,\text{fs}^3$ (*upper trace*), for $\alpha = -636\,\text{fs}^2$, $\beta = 500\,\text{fs}^3$ (*middle trace*) and for $\alpha = -636\,\text{fs}^2, \beta = 0\,\text{fs}^3$ (*lower trace*). Center wavelength is 810 nm, the FWHM of the shortest pulse is 45 fs

1.3.3 Control by Other Means of Systematic Pulse Shaping

We briefly recall some other approaches for systematic pulse shaping known and applied in the literature. First, we mention binary pulse shaping where successive phase jumps of π are written to a SLM [33, 34]. This approach has been successfully applied by Dela Cruz for distinguishing p-xylene and o-xylene structural isomers [35]. We note, that this distinction is also possible by means of systematic linear chirp [36].

We also mention femtosecond interferometry as a means for systematically shaping an effective laser field. Here, the original laser pulse is split into two identical replicas in a Michelson interferometer. Combining the two partial laser fields as a function of the length of the two arms leads to interference phenomena. Changing the delay time between the two arms of the interferometer then leads to a variation of the integrated laser intensity which in turn leads to a variation of the corresponding ion yields. If aligned with the necessary care the laser fields obtained by this interferometry allow to investigate in particular highly non-linear dynamical process, e.g. the formation of H_3^+ from ethane as shown in Fig. 1.10. Translating the movable mirror in steps of 10 nm then corresponds to a change in the delay time of about 60 as giving access to extremely high relative time resolution. Ultimately we expect that this approach might provide excess to a better understanding of the coupling of electron and nuclear dynamics [37, 38].

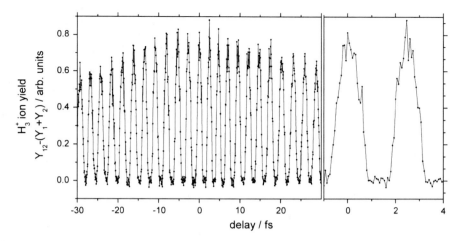

Fig. 1.10 Interferogramm for the formation of H_3^+ ions recorded at a chosen central wavelength of 800 nm and 15 μJ pulse energy in each arm of the interferometer, starting from unchirped pulses. Y_{12} is the interferometric signal; Y_1 and Y_2 are the signals for arm 1 and 2 at infinite temporal separation of the partial waves. Note, the time steps are 60 as. The stability of the interferometer is on the order of 120 as

1.3.4 Control by Genetic Algorithm

In previous sections we have investigated the role of linear and quadratic chirp on the yields of ions formed in the dissociative ionization of ethane. We now investigate the possible interplay of chirp contributions from different orders. In other words we search for the optimum spectral phase with respect to a given objective, e.g. maximizing a specific ion yield. This is a classical question for optimal control studies [39–41]. One of the central goals of the current investigation is the understanding and control of H_3^+ formation. Consequently, we first attempt to find the optimum laser field which maximizes the H_3^+ ion yield in the dissociative ionization of ethane. This goal is approached by using a 4f-shaper with a spatial light modulator for which the spectral phase is imprinted on 640 pixels. Thus, the phase shift for each pixel, i.e. each wavelength, can be varied independently and optimized by the genetic algorithm mentioned above (see experimental section). Figure 1.11 shows the ToF-MS of the H_3^+ ion region for the optimized laser field (upper trace) and for comparison the ToF-MS for an unchirped laser field at the same pulse energy (lower trace). In the case of the optimized laser field the total yield of the H_3^+ ion is about a factor of 8 higher than in the case of 45 fs pulses. This variation is considered a rather large value. More typical values reported in the literature are factors of 2 or 3 [39, 42, 43].

We now attempt to illustrate the characteristics of the underlying control scheme. To this end we show the FROG trace of the optimized laser field in Fig. 1.12a. We first note that the overall temporal width of the pulse is on the order of 130 fs, which matches with the findings of the systematic variation of the linear chirp at the

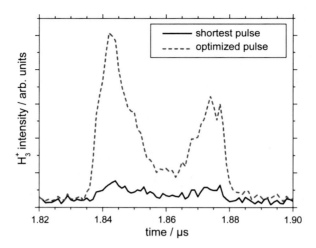

Fig. 1.11 H_3^+ TOF signal for the unchirped pulse (*black*), and for the genetic algorithm optimized pulse (*blue*). The pulse energy is 17 μJ/pulse

corresponding pulse energy [21]. Also, the spectral phase retrieved by the FROG algorithm suggests a predominance of a linear chirp. The SHG-FROG used in our experiment is capable of identifying a linear chirp, however it is not able to distinguish between a negative and positive chirp. Nevertheless we are able to characterize the pulses to the level desired by writing that spectral phase to the SLM which transforms the pulse found by optimization back into the shortest achievable laser pulse. In the case of the laser pulse presented in Fig. 1.12a writing a linear chirp with $\alpha = +900\,\mathrm{fs}^2$ and a quadratic chirp with $\beta = -3,333\,\mathrm{fs}^3$ to the SLM generates a close to 45 fs pulse depicted in Fig. 1.12c. Thus we conclude that the originally optimized laser pulse was dominated by linear and quadratic chirp components of $\alpha = -900\,\mathrm{fs}^2$ and $\beta = +3,333\,\mathrm{fs}^3$ respectively.

It is interesting to analyze the kinetic energy distribution of the H_3^+ ions formed with the optimized laser field. The procedure for this analysis was described in the experimental section. Figure 1.13 presents the kinetic energy distribution of the H_3^+ ions formed with the optimized laser field discussed above. Evidently the KED is spread out over more than 2 eV with the maximum probability occurring in the region of 3 eV. The overall KED shown here is very similar to the one reported by Kanya et al. [19], however the maximum probability occurred at slightly larger values there. Note that the KE's we discuss in this work refer to a specific fragment, here the H_3^+, while the total KE released into all fragments was reported in [19]. For the $C_2H_6^{++} \rightarrow H_3^+ + C_2H_3^+$ channel the ratio between the total KER and the KE(H_3^+) is 10 : 9. This small difference of 10 % does not fully account for the difference between the current KED and the one reported in [19]. At this point we cannot answer the question whether the KED depends on the fs-pulse shape. Further studies are required to resolve this issue.

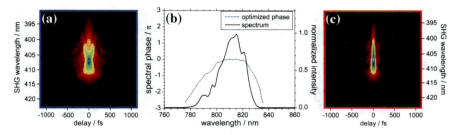

Fig. 1.12 a FROG of laser pulse optimized for H_3^+ (17 µJ), FWHM = 128 fs; **b** Spectrum and spectral phase retrieved from FROG; **c** FROG of the recompressed laser pulse obtained by adding a linear chirp with $\alpha = +900\,\text{fs}^2$ and a quadratic chirp with $\beta = -3{,}333\,\text{fs}^3$ to the pulse shown in (**a**). Hence, the main chirp components of the optimized pulse (**a**) are $\alpha = -900\,\text{fs}^2$, $\beta = +3{,}333\,\text{fs}^3$. The *vertical axis* of the FROG pictures is the SHG-wavelength in nm, the *horizontal axis* is the delay in fs

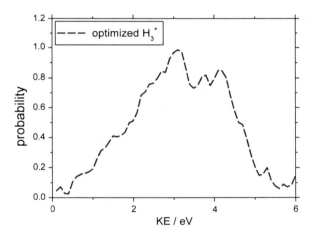

Fig. 1.13 Kinetic energy distribution of the H_3^+ ions formed via the optimized laser field

We have also performed GA studies for optimizing the H_3^+ ion yield at higher pulse energies. For a pulse energy of 80 µJ (not shown here) the optimal laser pulse found by the genetic algorithm has a pulse duration of about 60 fs. This is a little (but discernibly) longer than the duration of the shortest pulse achieved. Consequently a small linear chirp of $\alpha = 350\,\text{fs}^2$ has to be added to this pulse in order to recompress it to the shortest one. Therefore, the main chirp component of the optimized pulse is concluded to be $\alpha = -350\,\text{fs}^2$. This result matches the findings reported in [21]. With increasing pulse energy the dynamics become intensity driven and linear and quadratic chirp is not favoring high ion yields.

The comparison of results from systematically adding linear or quadratic chirp to a laser field with results from GA optimization demonstrates that systematic variation of chirp parameters is an efficient tool for maximizing ion yields for H_3^+ ions as well as for H^+ ions (the latter not shown here). The introduction of more complicated

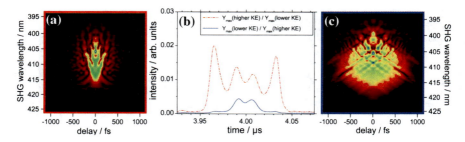

Fig. 1.14 a FROG of laser pulse optimized for Y_{max}(higher KE)/Y_{max}(lower KE) (82 μJ), FWHM = 116 fs; **b** CH_3^+ ion yields optimized for Y_{max}(higher KE)/Y_{max}(lower KE) (*red*) and Y_{max}(lower KE)/Y_{max}(higher KE) (*blue*); **c** FROG of laser pulse optimized for Y_{max}(lower KE)/Y_{max}(higher KE) (80 μJ) FWHM >400 fs. All information referring to the CH_3^+ ion. The *vertical axis* of the FROG pictures is the SHG-wavelength in nm, the *horizontal axis* is the delay in fs

spectral phases with a genetic algorithm does not cause significantly better results than already obtained with systematic chirp.

In Sect. 1.3.1 we have mentioned the ability to control the ratio of fast CH_3^+ ions (originating from the dicationic state) to slow CH_3^+ ions originating from the monocationic state of ethane. In the following we search for the optimum laser field for maximizing the ratio of ion yields of slow to fast CH_3^+ ions, (Y_{max}(lower KE)/Y_{max}(higher KE), and vice versa. Since the peaks for fast and slow components of CH_3^+ are not always well separated, integration of ion yields is not the method of choice. Instead we analyze the maximum signal amplitudes at the respective peak positions.

Figure 1.14a, c show FROG traces obtained for optimizing the ion yield ratios Y_{max}(higher KE)/Y_{max}(lower KE) and Y_{max}(lower KE)/Y_{max}(higher KE) respectively. The corresponding ion yields themselves are shown in Fig. 1.14b. Overall, the ion yield ratio Y_{max}(higher KE)/Y_{max}(lower KE) can be changed from about 1 : 5 to 2 : 1.

Further analysis of the FROG traces shows that the laser pulse maximizing the ratio Y_{max}(higher KE)/Y_{max}(lower KE) (Fig. 1.14a) can be recompressed from about 116 fs to 52 fs by mainly adding a negative linear chirp of $\alpha = -550\,fs^2$. Other more complex phase components are present but of minor importance. For the laser pulse depicted in Fig. 1.14c, i.e. the one optimizing the ratio Y_{max}(lower KE)/Y_{max}(higher KE) we do not find a simple spectral phase characteristic. We note that the overall temporal pulse width is in this case larger than 400 fs. Since both optimizations were conducted at a pulse energy of about 80 μJ/pulse the smaller temporal width of the pulse in Fig. 1.14a implies a considerably higher intensity compared to the pulse in Fig. 1.14c. The higher laser intensity ultimately favors the fast component of the CH_3^+ ions, which originate from doubly ionized ethane ions. This adds to a self-consistent picture. The laser field in Fig. 1.14c is very complex. We note that in the maximization of the ratio Y_{max}(lower KE)/Y_{max}(higher KE) effectively

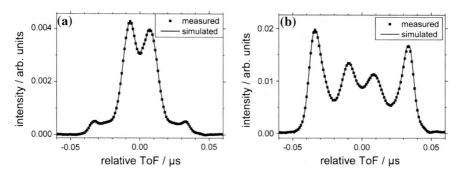

Fig. 1.15 Optimal control of CH_3^+ ion yields. **a** *Left graph* ToF-MS for a laser field maximizing the ratio Y_{max}(lower KE)/Y_{max}(higher KE). **b** *Right graph* ToF-MS for a laser field maximizing the ratio Y_{max}(higher KE)/Y_{max}(lower KE)

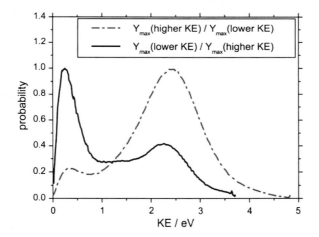

Fig. 1.16 Kinetic energy distributions for optimized laser fields: *solid line* maximization of the ratio Y_{max}(lower KE)/Y_{max}(higher KE). *dotted line* maximization of the ratio Y_{max}(higher KE)/Y_{max}(lower KE)

the total yield of both fast and slow ions is decreased, but the fast component is even stronger suppressed than the slow component.

The ToF spectra (recorded at a pressure of 3.9×10^{-6} mbar) shown in Fig. 1.14b evidently involve very different KE distributions of the CH_3^+ ions. In order to quantify this observation we have again performed a KE analysis. Figure 1.15 shows the relevant experimental ToF spectra together with the best simulations, which is based on the KE distributions displayed in Fig. 1.16.

The two KED's shown in Fig. 1.16 both exhibit probability maxima at about 0.4 eV and at 2.5 eV. However the relative peak probability differs significantly. For optimization of the slow versus fast component the ratio of peak values is 1:0.4, while for optimization of the fast versus slow component it is 0.2:1. While previous analysis only extracted average KER from the ToF spectra we have arrived at a more detailed picture here by presenting the entire KED's.

1.4 Summary and Conclusions

We have presented a systematic investigation of the influence of pulse shaping on the formation of fragment ions in the dissociative ionization of ethane employing intense ultra-short laser pulses. We have demonstrated that concepts aiming at understanding and those aiming at controlling go hand in hand. E.g., previous work directed at understanding the dynamics had provided evidence for hydrogen scrambling within the dicationic state of ethane. Additional experiments with the goal of controlling the yields of fragment ions as well as their properties (here the kinetic energy) have shed additional light on the mechanisms involved. In particular we see clear evidence of intra-charge-state control as well as inter-charge-state control. This opens access to improved understanding of the competition between single ionization and double ionization. Ultimately it is hoped that experimental work as outlined here stimulates further theoretical investigation in order to better understand the electron and nuclear dynamics involved in the dissociative ionization of small hydrocarbon molecules.

Acknowledgments Contributions from G. Urbasch, M. Galbreith, and K. Bücker are gratefully acknowledged. N.R. acknowledges financial support from the Fonds der Chemischen Industrie.

References

1. A. Giusti-Suzor, F.H. Mies, L.F. DiMauro, E. Charron, B. Yang, J. Phys. B: At. Mol. Opt. Phys. **28**, 309 (1995)
2. B. Sheehy, L.F. DiMauro, Annu. Rev. Phys. Chem. **47**, 463 (1996)
3. S.A. Buzza, E.M. Snyder, D.A. Card, D.E. Folmer, A.W. Castleman, J. Chem. Phys. **105**, 7425 (1996)
4. M.J. Dewitt, D.W. Peters, R.J. Levis, Chem. Phys. **218**, 211 (1997)
5. J.H. Posthumus, Rep. Prog. Phys. **67**, 623 (2004)
6. I.V. Hertel, W. Radloff, Rep. Prog. Phys. **69**, 1897 (2006)
7. A. Vredenborg, C.S. Lehmann, D. Irimia, W.G. Roeterdink, M.H.M. Janssen, Chemphyschem **12**, 1459 (2011)
8. Q.L. Liu, J.K. Wang, A.H. Zewail, Nature **364**, 427 (1993)
9. H. Arnolds, M. Bonn, Surf. Sci. Rep. **65**, 45 (2010)
10. R. Itakura, P. Liu, Y. Furukawa, T. Okino, K. Yamanouchi, and H. Nakano, J. Chem. Phys. **127**, 104306 (2007)
11. C.Y. Wu, Q.Q. Liang, M. Liu, Y.K. Deng, Q.H. Gong, Phys. Rev. A **75**, 043408 (2007)
12. K. Hoshina, Y. Furukawa, T. Okino, K. Yamanouchi, J Chem Phys **129**, 104302 (2008)
13. P.M. Kraus, M.C. Schwarzer, N. Schirmel, G. Urbasch, G. Frenking, K.-M. Weitzel, J. Chem. Phys. **134**, 114302 (2011)
14. H. Conroy, J. Chem. Phys. **40**, 603 (1964)
15. W. Koch, G. Frenking, J. Gauss, D. Cremer, J. Am. Chem. Soc. **108**, 5808 (1986)
16. M.D. Burrows, S.R. Ryan, W.E. Lamb, L.C. Mcintyre, J. Chem. Phys. **71**, 4931 (1979)

17. T. Okino, Y. Furukawa, P. Liu, T. Ichikawa, R. Itakura, K. Hoshina, K. Yamanouchi, H. Nakano, Chem. Phys. Lett. **419**, 223 (2006)
18. T. Okino, Y. Furukawa, P. Liu, T. Ichikawa, R. Itakura, K. Hoshina, K. Yamanouchi, H. Nakano, Chem. Phys. Lett. **423**, 220 (2006)
19. R. Kanya, T. Kudou, N. Schirmel, S. Miura, K.-M. Weitzel, K. Hoshina, K. Yamanouchi, J. Chem. Phys. **136**, 204309 (2012)
20. J.H.D. Eland, Rapid Commun. Mass Spectrom. **10**, 1560 (1996)
21. N. Schirmel, N. Reusch, P. Horsch, K.-M. Weitzel, Faraday Discuss. **163**, 461–474 (2013)
22. P. Balling, D.J. Maas, L.D. Noordam, Phys. Rev. A **50**, 4276 (1994)
23. R. Trebino, in *Frequency-Resolved Optical Gating: The Measurement of Ultrashort Laser Pulses* (Kluwer Academic Publishers, Norwell, 2000)
24. D.E. Goldberg, in *Genetic Algorithms in Search, Optimization, and Machine Learning* (Addison-Wesley, Reading, UK, 1993)
25. W.C. Wiley, I.H. McLaren, Rev. Sci. Instrum. **26**, 1150 (1955)
26. R. Stockbauer, Int. J. Mass Spectrom. Ion Phys. **25**, 89 (1977)
27. J.L. Franklin, P.M. Hierl, D.A. Whan, J. Chem. Phys. **47**, 3148–3153 (1967)
28. K.-M. Weitzel, F. Güthe, J. Mähnert, R. Locht, H. Baumgärtel, Chem. Phys. **201**, 287 (1995)
29. R. Itakura, K. Yamanouchi, T. Tanabe, T. Okamoto, F. Kannari, J. Chem. Phys. **119**, 4179 (2003)
30. D. Mathur, F.A. Rajgara, J. Chem. Phys. **120**, 5616 (2004)
31. V.V. Lozovoy, X. Zhu, T.C. Gunaratne, D.A. Harris, J.C. Shane, M. Dantus, J. Phys. Chem. A **112**, 3789 (2008)
32. M. Wollenhaupt, L. Englert, A. Horn, T. Baumert, J. Laser Micro Nanoeng. **4**, 144 (2009)
33. M. Comstock, V.V. Lozovoy, I. Pastirk, M. Dantus, Opt. Express **12**, 1061 (2004)
34. V.V. Lozovoy, T.C. Gunaratne, J.C. Shane, M. Dantus, Chemphyschem **7**, 2471 (2006)
35. J.M. Dela Cruz, V.V. Lozovoy, M. Dantus, J. Phys. Chem. A **109**, 8447 (2005)
36. G. Urbasch, H.G. Breunig, K.M. Weitzel, Chemphyschem **8**, 2185 (2007)
37. H.G. Breunig, G. Urbasch, K.M. Weitzel, J. Chem. Phys. **128**, 121101 (2008)
38. M.V. Korolkov, K.M. Weitzel, Chem. Phys. Lett. **487**, 209 (2010)
39. A. Assion, T. Baumert, M. Bergt, T. Brixner, B. Kiefer, V. Seyfried, M. Strehle, G. Gerber, Science **282**, 919 (1998)
40. R. Kosloff, S.A. Rice, P. Gaspard, S. Tersigni, D.J. Tannor, Chem. Phys. **139**, 201 (1989)
41. S.A. Rice, M.S. Zhao, in *Optical Control of Molecular Dynamics* (Wiley, New York, 2000)
42. M. Bergt, T. Brixner, B. Kiefer, M. Strehle, G. Gerber, J. Phys. Chem. A **103**, 10381 (1999)
43. H.G. Breunig, A. Lauer, K.M. Weitzel, J. Phys. Chem. A **110**, 6395 (2006)

Chapter 2
Electron and Ion Coincidence Momentum Imaging of Multichannel Dissociative Ionization of Ethanol in Intense Laser Fields

Ryuji Itakura, Kouichi Hosaka, Atsushi Yokoyama, Tomoya Ikuta, Fumihiko Kannari and Kaoru Yamanouchi

Abstract We investigate the multichannel dissociative ionization of C_2H_5OH in intense laser fields by the photoelectron-photoion coincidence momentum imaging and identify separately the ionization and subsequent electronic excitation in C_2H_5OH. From the energy correlation between a photoelectron and a fragment ion, we reveal the amount of the internal energy gained by $C_2H_5OH^+$ from the laser field varies depending on the respective ionization and electronic excitation pathways. It is found that $C_2H_5OH^+$ prepared in the electronic ground state associated with the moment of the photoelectron emission gains larger internal energy at the end of the laser pulse than that prepared in the electronically excited state at the moment of the photoelectron emission.

2.1 Introduction

When molecules are exposed to an intense laser field whose strength is comparable with the Coulombic electric field binding the valence electrons within the molecules, electronic excitation and ionization proceed, leading to the decomposition of molecules into atomic and molecular fragment cations [1, 2]. Electronic

---

R. Itakura (✉) · K. Hosaka · A. Yokoyama · T. Ikuta · K. Yamanouchi
Quantum Beam Science Directorate, Kansai Photon Science Institute, Japan Atomic
Energy Agency, 8-1-7 Umemidai Kizugawa, Kyoto 619-0215, Japan
e-mail: itakura.ryuji@jaea.go.jp

T. Ikuta · F. Kannari
Department of Electronics and Electrical Engineering, Keio University, 3-14-1 Hiyoshi,
Kouhoku-ku, Yokohama 223-8522, Japan
e-mail: kannari@elec.keio.ac.jp

K. Yamanouchi
Department of Chemistry, School of Science, The University of Tokyo, 7-3-1 Hongo
Bunkyo-ku, Tokyo 113-0033, Japan
e-mail: kaoru@chem.s.u-tokyo.ac.jp

K. Yamanouchi et al. (eds.), *Progress in Ultrafast Intense Laser Science XI*,
Springer Series in Chemical Physics 109, DOI: 10.1007/978-3-319-06731-5_2,
© Springer International Publishing Switzerland 2015

excitation, ionization, and dissociation pathways have been discussed since the pioneering work on the multiphoton ionization of molecules with an intense laser field [3]. For instance, the fragment molecular ions from C_2H_4 in intense laser fields were expected to be formed through the removal of an inner valence electron based on the comparison between the intensity dependence of product ion yields and the calculation of multichannel coupled rate equations [4]. Theoretically, the electronic excitation of molecules in intense laser fields has been discussed in terms of nonadiabatic transition between electronic states [5–8], while ionization has been described using multiphoton ionization or tunneling ionizaiton [9–14] separately from the excitation of a bound electron.

Recently, measurements of high-order harmonic spectra for spatially aligned molecules have been reported. Combined with the theoretical simulation of multichannel electron wavepacket propagation, the formation of the electronically excited state of molecules at the moment of photoelectron emission has been actively discussed [15–17]. As another experimental approach to investigate the electronic excitation dynamics associated with ionization, pump-probe experiments for monitoring the product ion yields were also reported [18–21]. The oscillating frequency of the observed ion yields as a function of the pump-probe delay was compared with the vibrational frequencies of the potential energy surfaces, indicating the populations in the electronically excited states of ions [18, 19]. With the knowledge of the correlation between specific electronic states of a parent ion and its product ion species, the detection of fragment ions was regarded as evidence of the formation of the electronically excited states of the parent ion [20, 21]. The yield ratios of the parent and fragment ions were compared with the results of the time-dependent multichannel calculation of the ionization process [22, 23].

However, in order to discuss the complex correlation between ionization and subsequent electronic excitation within an intense laser field, further investigation is required. In particular, it is unclear when the electronically excited ions are prepared, that is, at the moment of photoelectron emission or after the electron emission. It is also important to reveal the amount of internal energy that can be gained by ions in laser fields in the respective steps of the ionization and electronic excitation processes. In order to answer these questions, we have performed coincidence detection of a photoelectron and a product ion, and have investigated the correlation between photoelectron emission and subsequent dissociation channels.

The advantages of the photoelectron-photoion coincidence (PEPICO) measurement have been demonstrated in the investigation of dissociative ionization dynamics induced by a variety of light sources such as a He lamp [24] and synchrotron radiation [25–27]. In the weak field regime, dissociation dynamics of molecules in their electronically excited neutral states have been investigated by PEPICO measurement with a femtosecond pump-probe setup [28, 29]. The PEPICO technique can also provide rich information on dissociative ionization processes of small polyatomic molecules in intense laser fields whose intensity is strong enough for observing above threshold ionization [23, 30–36]. In this chapter, we would like to show how we can extract the characteristic aspects of the multichannel dissociative ionization

2 Electron and Ion Coincidence Momentum Imaging

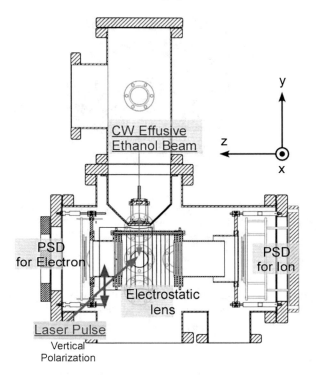

Fig. 2.1 Photoelectron-photoion coincidence momentum imaging apparatus

dynamics of C_2H_5OH in intense laser fields through the PEPICO momentum imaging by referring to our recent related studies [33, 35, 36].

2.2 Photoelectron-Photoion Coincidence Momentum Imaging

Our PEPICO apparatus is composed of (i) a continuous effusive beam source for introducing a sample gas, (ii) a set of electrostatic lenses for guiding photoelectrons and photoions, and (iii) two sets of fast position-sensitive detectors (PSD) for recording three dimensional momentum vectors of photoelectrons and photoions, as illustrated in Fig. 2.1.

To secure the coincidence condition, the ultrahigh vacuum condition in the experimental chamber is necessary. Background molecules remaining at the focal spot of a laser pulse can easily be ionized in intense laser fields, causing false coincidence events. Therefore, we kept the base pressure in our apparatus below 10^{-8} Pa. The sample gas was continuously introduced into the main experimental chamber as an effusive molecular beam through a micro-syringe and a skimmer. The translational temperatures of the effusive beam in the directions perpendicular to and parallel to

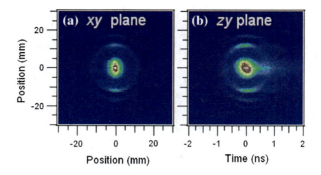

Fig. 2.2 Raw images of the photoelectrons for the channel yielding $C_2H_5OH^+$ when $\lambda \sim 783$ nm, $\tau = 35$ fs, and $I_0 = 9$ TW/cm^2. Images were obtained through the projection on **a** the PSD plane (xy) and **b** the zy plane, in which the horizontal axis (z) is along the time-of-flight axis. The vertical axis (y) in both images is parallel to the direction of the laser polarization

the molecular beam axis are estimated to be around 6 and 100 K, respectively, from the observed momentum distribution of the parent ion, $C_2H_5OH^+$. In order to keep the coincidence condition, the density of the sample molecules at the focal point was set to be sufficiently low. The density of the molecular beam was controlled with a variable leak valve, and the count rate of the detection was set to be smaller than 0.3 events per laser shot.

A stack of electrode plates provide the static electric fields so that photoelectrons and photoions produced at the laser focal spot are accelerated towards the respective PSDs, and are temporally and spatially focused on the PSDs [26, 37]. In this configuration with the acceleration voltage of 3 kV, the maximum energy of photoelectrons that can be detected with the detection solid angle of 4π sr is around 20 eV.

The fast PSDs composed of microchannel plates (MCPs) and delay-line anodes (RoentDek Hex80) allow us to determine three dimensional velocities of the respective photoelectrons and photoions from their respective two-dimensional (2D) arrival positions and flight time. It should be emphasized that the resolution of the photoelectron momentum along the time-of-flight axis is sufficiently high thanks to the high temporal resolution (25 ps/bin) of the time-to-digital converters (RoentDek, TDC8HP), to which the timing signals from the MCP and delay line anodes were sent through the fast amplifier and constant fraction discriminator (RoentDek, ATR19). Figure 2.2 shows that the observed photoelectron image projected on the zy plane containing the time-of-flight (z) axis and the laser polarization direction (y axis) is very close to that projected on the PSD plane (xy plane). Thus, the 2D position and flight-time can be converted to the three-dimensional momenta of photoelectrons as well as to those of photoions.

The PEPICO measurement was carried out using two different laser wavelengths, that is, the fundamental near-infrared (NIR) output ($\lambda \sim 783$ nm) of a Ti:Sapphire regenerative amplifier and the frequency-doubled ultraviolet (UV) output

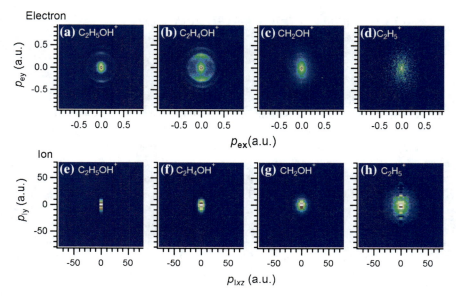

Fig. 2.3 2D sliced images of 3D momentum distributions of photoelectrons for the channels yielding **a** $C_2H_5OH^+$, **b** $C_2H_4OH^+$, **c** CH_2OH^+, and **d** $C_2H_5^+$, and 2D cuts of the 3D momentum distributions of **e** $C_2H_5OH^+$, **f** $C_2H_4OH^+$, **g** CH_2OH^+, and **h** $C_2H_5^+$ when $\lambda = 783$ nm, $I_0 = 9$ TWcm^2, and $\tau = 35$ fs. The vertical (y) axis is along the laser polarization direction. The unit of the coordinates is the atomic unit (a.u.) in both electron and ion images. The sliced images in (**a**)–(**d**) are plotted for the momenta with $|p_{ez}| < 0.05$ a.u. p_{Ixz} is defined as $|p_{Ixz}| = (p_{Ix}^2 + p_{Iz}^2)^{1/2}$, and the image in the area of $p_{Ixz} < 0$ is a mirror image of that in the area of $p_{Ixz} > 0$. The x,y, z coordinates are illustrated in Fig. 2.1

($\lambda \sim 400$ nm). The duration of the transform limited NIR pulses was measured to be $\tau = 35$ fs with the second harmonic generation frequency-resolved optical gating (FROG) [38]. A zero-order half wave plate and a thin film polarizer were used for changing the laser intensity. The pulse duration τ was controlled through the adjustment of the distance between a pair of gratings in the compressor in the chirped pulse amplification system. The UV pulse was generated through the frequency doubling with a β barium-borate (type-I) crystal of 0.1 mm thickness. The pulse duration of the UV pulse was measured to be $\tau = 96$ fs with the self-diffraction FROG [38]. The intensity of the UV pulse was adjusted by changing the diameter of the UV beam with an iris. The laser pulses were focused on the effusive ethanol beam. The laser peak intensity I_0 at the focal spot was estimated using the ponderomotive energy shift of the peaks appearing in the photoelectron spectra recorded for the channel yielding $C_2H_5OH^+$.

The 2D cuts of the 3D momentum distributions of the photoelectrons are shown in Fig. 2.3a–d for the channels yielding $C_2H_5OH^+$, $C_2H_4OH^+$, CH_2OH^+, and $C_2H_5^+$, when $\lambda \sim 783$ nm, $\tau = 35$ fs, and $I_0 = 9$ TW/cm^2. The 3D momentum distributions of the ions are also shown in Fig. 2.3e–h.

2.3 Channel-Specific Photoelectron Spectra

2.3.1 Electronic Energy Levels of $C_2H_5OH^+$ and Appearance Energies of Product Ions

From the 3D momentum distributions of the photoelectrons, we can straightforwardly obtain the photoelectron spectra for the respective channels. The channel-specific photoelectron spectra reveal the electronic states prepared at the moment of photoelectron emission as well as the final state of $C_2H_5OH^+$ correlated with the product ion formation. The energy diagram of the electronic states of $C_2H_5OH^+$ and the threshold energy for the respective channels producing fragment cations are illustrated in Fig. 2.4, where the vertical energies of the electronic states of $C_2H_5OH^+$ and the appearance energies of the product ions are shown. From the photoelectron spectrum of C_2H_5OH [39] recorded using a He(I) ($\lambda = 58.4$ nm) light source and the yield ratios of the product ions [40] obtained as a function of the internal energy of $C_2H_5OH^+$, it was shown that $C_2H_5OH^+$ is formed as a final product only when $C_2H_5OH^+$ is populated in the electronic ground state and that $C_2H_4OH^+$ is produced from the vibrationally excited states of $C_2H_5OH^+$ whose energy is larger than the threshold energy of ~ 0.3 eV in the electronic ground state. When $C_2H_5OH^+$ is prepared in the electronically excited states, dissociation into CH_2OH^+, $C_2H_4OH^+$, $C_2H_5^+$, $C_2H_3^+$, and CH_3^+ proceeds [39, 40]. Therefore, from the relative yields of these fragment ions detected in our PEPICO measurements, the final relative yield of $C_2H_5OH^+$ in the electronically excited states after the interaction with intense laser fields can be determined. As illustrated in Fig. 2.4, (i) CH_2OH^+ is produced from the first or the higher-lying electronically excited state of $C_2H_5OH^+$, (ii) $C_2H_5^+$, $C_2H_3^+$ and CH_3^+ are produced when $C_2H_5OH^+$ is prepared in the second or the higher-lying electronically excited states, and $C_2H_4OH^+$ is produced from the electronically excited states as well as from the vibrationally excited states in the electronic ground state of $C_2H_5OH^+$ [40].

2.3.2 Near-Infrared Laser Fields

2.3.2.1 Ionization and Electronic Excitation Pathways in Near-Infrared Fields

The channel-specific photoelectron spectra when $\lambda \sim 783$ nm and $\tau = 35$ fs are shown in Fig. 2.5a–c for the three different laser intensities $I_0 = 9$, 17, and 23 TW/cm^2 [36]. At $I_0 = 9$ TW/cm^2, a series of distinct peaks ascribed to the above-threshold-ionization (ATI) processes were observed for the channels yielding $C_2H_5OH^+$ and $C_2H_4OH^+$. These ATI peaks for the channel yielding $C_2H_4OH^+$ are shifted to the lower energy side than those for the channel yielding $C_2H_5OH^+$ by around 0.3 eV, corresponding to the vibrational excitation energy required for the H elimination [40].

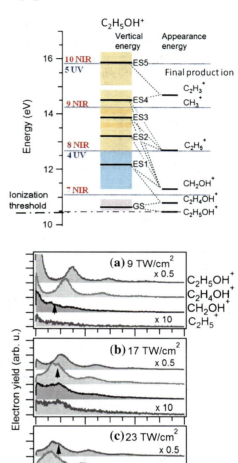

Fig. 2.4 Energy levels of electronic states of $C_2H_5OH^+$ and appearance energies of the respective product ions. The origin of the energy (*left*) axis corresponds to the electronic and vibrational ground state of neutral C_2H_5OH

Fig. 2.5 Photoelectron spectra for the formation of $C_2H_5OH^+$, $C_2H_4OH^+$, CH_2OH^+, and $C_2H_5^+$ (from the *top*) at the laser peak intensities of $I_0 =$ **a** 9, **b** 17, and **c** 23 TW/cm^2 with $\lambda \sim$ 783 nm, $\tau = 35$ fs. The spectra for the $C_2H_5OH^+$ and $C_2H_5^+$ formation are multiplied by 0.5 and 10, respectively. Peaks at around 0.9 eV at which arrows point are assigned to Freeman resonances

In contrast, in the photoelectron spectra for the channels yielding CH_2OH^+ and $C_2H_5^+$, a diffuse spectral feature with no distinct peak was observed, indicating that the first electronically excited state of $C_2H_5OH^+$ is prepared at the same time as a photoelectron is emitted. This interpretation is based on the previous experimental results that the photoelectron spectrum of C_2H_5OH exhibits a broad spectral feature in the spectral region corresponding to the formation of the electronically excited state of $C_2H_5OH^+$ [41]. This broad spectral feature is ascribed to the Franck-Condon overlap between the electronic ground state of neutral C_2H_5OH and the first electronically excited state of $C_2H_5OH^+$. Considering that the yield of the electronically excited states of $C_2H_5OH^+$ represented by the broad spectral feature is comparable to the yield of the electronic ground state, the excitation mechanisms are not likely to be the recollision [42, 43] or shake-up [44] processes through which the electronic excitation in molecular ions is expected to proceed with only small probabilities.

2.3.2.2 Laser Intensity Dependence

As I_0 increases from 9 to 23 TW/cm^2 while keeping $\tau = 35$ fs, the differences in the spectral feature among the channels yielding $C_2H_5OH^+$, $C_2H_4OH^+$, CH_2OH^+, and $C_2H_5^+$ becomes less prominent, and a series of the ATI peaks become distinct in the channels yielding CH_2OH^+ and $C_2H_5^+$ as shown in Fig. 2.5b–c. Indeed, as I_0 increases, the photoelectron spectra for the channels yielding CH_2OH^+ and $C_2H_5^+$ become closer to those for the channels yielding $C_2H_5OH^+$ and $C_2H_4OH^+$ produced from the electronic ground state of $C_2H_5OH^+$, indicating that the electronic ground state of $C_2H_5OH^+$ tends to be formed preferentially at the moment of the photoelectron emission even in these fragment channels.

Given that the electronic ground state of $C_2H_5OH^+$ is prepared at the moment of photoelectron emission, additional electronic excitation as a post-ionization process is required energetically for yielding CH_2OH^+ and $C_2H_5^+$. As I_0 increases the population of $C_2H_5OH^+$ in the electronic ground state at the moment of photoelectron emission becomes dominant for the channels yielding CH_2OH^+ and $C_2H_5^+$. This means that the increase of the yield ratios of the fragment ions such as CH_2OH^+ and $C_2H_5^+$ is ascribable to the enhancement of the electronic excitation from the electronic ground state of $C_2H_5OH^+$ within the laser field.

2.3.2.3 Pulse Duration Dependence

As the laser pulse duration increases to $\tau = 200$ and 800 fs while the peak intensity I_0 is being kept in the range between 17 and 19 TW/cm^2, the channel-specific photoelectron spectra were recorded as shown in Fig. 2.6. The photoelectron spectra in Fig. 2.6a, b were obtained with the positively chirped 200 and 800 fs pulses, respectively, while those in Fig. 2.6c, d were obtained with the negatively chirped 200 and 800 fs pulses, respectively. The observation that (i) the ATI peaks become prominent in the spectra for the channels yielding CH_2OH^+ and $C_2H_5^+$ as τ increases from 200 to 800 fs irrespective of the direction of the chirp and (ii) their spectral features become closer to those for the channels yielding $C_2H_5OH^+$ and $C_2H_4OH^+$ indicates that the electronic excitation of $C_2H_5OH^+$ after the photoelectron emission is enhanced as the laser pulse duration increases [33, 36].

As described in Sect. 2.3.2.2, this enhancement of the electronic excitation after ionization is similarly exhibited when the laser intensity I_0 increases. The enhancement by stretching the pulse duration may be ascribed to the evolution of the vibrational wavepacket on the potential energy surface (PES) of the electronic ground state

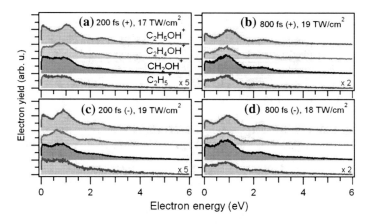

Fig. 2.6 Photoelectron spectra for the formation of $C_2H_5OH^+$, $C_2H_4OH^+$, CH_2OH^+, and $C_2H_5^+$ (from the *top*) at laser peak intensities in the range between $I_0 = 17$ and 19 TW/cm^2. The laser pulse ($\lambda \sim 783$ nm) is positively chirped with $\tau =$ **a** 200 and **b** 800 fs, and negatively chirped with $\tau =$ **c** 200 and **d** 800 fs

of $C_2H_5OH^+$. When the pulse duration is long enough, the wavepacket is expected to reach the region on the PES where non-adiabatic couplings with other electronic states are efficiently induced, and consequently, the electronic excitation is enhanced as demonstrated theoretically [7, 8].

Both of the photoelectron energy spectra correlated with CH_2OH^+ and $C_2H_5^+$ were found to be synthesized by a linear combination of the spectra correlated with $C_2H_4OH^+$ and $C_2H_5OH^+$ [33]. The ratio of the coefficients in the linear combination of the spectra correlated with $C_2H_4OH^+$ and $C_2H_5OH^+$, which are normalized so that the integrations over the respective spectral areas become unity, are evaluated through least-square fits to be 0.76(1) and 0.24(1) when $\tau = 800$ fs and $I_0 \sim 19$ TW/cm^2. The fact that the spectra for the channels yielding CH_2OH^+ and $C_2H_5^+$ are reproduced well by the linear combination indicates that (i) the electronic excitation occurs mostly after the electronic ground state of $C_2H_5OH^+$ is prepared at the moment of the photoelectron emission, and (ii) the probability of the formation of the electronically excited states at the moment of the photoelectron emission is negligibly small when $I_0 \sim 19$ TW/cm^2 and $\tau = 800$ fs.

2.3.2.4 Freeman Resonance

It should be noted that a broad peak can be identified in the photoelectron spectrum at the electron energy of $E_{elec} \sim 0.9$ eV for the channel yielding CH_2OH^+ at $I_0 = 9$ TW/cm^2 and $\tau = 35$ fs as shown in Fig. 2.5a. Similar broad peaks at $E_{elec} \sim 0.9$ eV can also be recognized in the photoelectron spectra for the channel yielding $C_2H_4OH^+$ at $I_0 = 17$ TW/cm^2 in Fig. 2.5b and for that yielding $C_2H_5OH^+$ at $I_0 = 23$ TW/cm^2 in Fig. 2.5c. These peaks can be ascribed to the Freeman resonance [45] with the 3 s Rydberg levels of neutral C_2H_5OH [46], whose level energy is in good agreement with the observed photoelectron energy of 0.9 eV. The laser intensities at which a peak of the Freeman resonance appears are different from each other in the channels yielding CH_2OH^+, $C_2H_4OH^+$, and $C_2H_5OH^+$, while the observed photoelectron energies are almost the same. The resonant Rydberg states in the respective ionization pathways may have different ion cores corresponding to the ionic states prepared at the moment of the photoelectron emission. The nearly equal values of the photoelectron energies for the resonance suggest that the resonant Rydberg levels in these ionization pathways are all 3 s levels with different electronic or vibrational states of the ion core.

2.3.3 Ultraviolet Laser Fields

2.3.3.1 Ionization and Electronic Excitation Pathways in Ultraviolet Fields

When the wavelength of laser pulses is in the ultraviolet (UV) region, multiphoton ionization is preferred as an ionization mechanism rather than tunneling ionization. Practically, the series of ATI peaks in the photoelectron spectra can be separated well because of the larger photon energy than the peak width of ATI peaks. Therefore, the decomposition of the structured peak profile into multiple components becomes possible as described below. The channel specific photoelectron spectra measured with an intense UV field ($\lambda \sim 400$ nm) at $I_0 = 1.5$ and 17 TW/cm^2 are shown in Fig. 2.7a, b, respectively, for the channels yielding $C_2H_5OH^+$, $C_2H_4OH^+$, CH_2OH^+, and $C_2H_5^+$ [35].

The spectrum for the $C_2H_5OH^+$ formation in Fig. 2.7a exhibits a relatively sharp peak at 1.74 eV corresponding to the four-photon ionization into the electronic ground state of $C_2H_5OH^+$. As I_0 increases from 1.5 to 17 TW/cm^2, a peak assignable to the five-photon ATI forming the electronic ground state of $C_2H_5OH^+$ is also identified at ~ 4.7 eV as shown in Fig. 2.7b.

In the channel-specific photoelectron spectrum for the channel yielding CH_2OH^+ at $I_0 = 1.5$ TW/cm^2 in Fig. 2.7a, a broad peak at 0.5 eV is assigned to the four-photon ionization into the first electronically excited state of $C_2H_5OH^+$. This broad spectral feature represents the wide Franck-Condon overlap between the electronic ground state of neutral C_2H_5OH and the first electronically excited state of $C_2H_5OH^+$ as observed in the previous photoelectron spectrum [39]. As I_0 increases to 17 TW/cm^2, not only the ionization to the first electronically excited state, but also the ionization to the electronic ground state appears as a peak at 1.47 eV as shown in Fig. 2.7b.

In the spectra for the channels yielding $C_2H_4OH^+$ and $C_2H_5^+$ in Fig. 2.7a, b, there are two components assignable to the electronic ground state and the first electronically excited state of $C_2H_5OH^+$ at $I_0 = 1.5$ and 17 TW/cm^2, respectively.

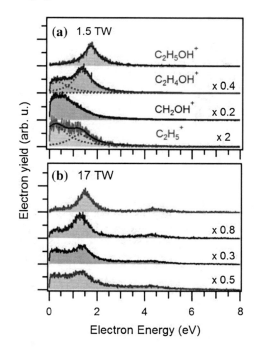

Fig. 2.7 Photoelectron spectra for the formation of $C_2H_5OH^+$, $C_2H_4OH^+$, CH_2OH^+, and $C_2H_5^+$ (from the *top*) at the laser peak intensities of $I_0 = $ **a** 1.5 and **b** 17 TW/cm^2 with $\lambda \sim 400$ nm, $\tau = 96$ fs. The *solid curves* are fitting results, which consist of several Lorentzian curves shown as the *dotted curves*. See the details in the text

2.3.3.2 Decomposition of Photoelectron Spectra

The channel-specific spectra for the channels yielding $C_2H_5OH^+$, CH_2OH^+, $C_2H_4OH^+$, $C_2H_5^+$, $C_2H_3^+$, and CH_3^+ are decomposed into Lorentzian functions by a least-squares analysis, and the nascent relative yields of the electronic ground state and the first electronically excited state of $C_2H_5OH^+$ prepared at the moment of the photoelectron emission are obtained from the cross-section area of the respective Lorentzian components [35]. Hereafter, the nascent relative yields of the ground state and the first excited state of $C_2H_5OH^+$ at the laser peak intensity I_0 are denoted as $Y_i^G(I_0)$ and $Y_i^E(I_0)$, respectively, where $Y_i^G(I_0) + Y_i^E(I_0) = 1$. By the sum of the respective contributions from the six channels yielding $C_2H_5OH^+$, CH_2OH^+, $C_2H_4OH^+$, $C_2H_5^+$, $C_2H_3^+$, and CH_3^+, the nascent relative yields are determined to be $Y_i^G(I_0) = 0.32$ and $Y_i^E(I_0) = 0.68$ at $I_0 = 1.5$ TW/cm^2.

By the subsequent interaction with the laser field, the electronic ground state of $C_2H_5OH^+$ can be excited into the higher-lying electronically excited state. The final relative yields of the electronic ground state and the electronically excited states of $C_2H_5OH^+$ are denoted as $Y_f^G(I_0)$ and $Y_f^E(I_0)$, respectively, where $Y_f^G(I_0) + Y_f^E(I_0) = 1$. From the correlation between the product ions and the electronic states of $C_2H_5OH^+$ illustrated in Fig. 2.4, the final relative yield of the electronic ground state of $C_2H_5OH^+$ is determined to be $Y_f^G(I_0) = 0.26$ at $I_0 = 1.5$ TW/cm^2. After the electronic excitation subsequent to the photoelectron emission, the population is found to be transferred not only to the first electronically excited state but also

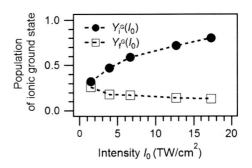

Fig. 2.8 Population ratio of the electronic ground state of $C_2H_5OH^+$ **a** at the moment of photoelectron emission ($Y_i^G(I_0)$) and **b** after interaction with a laser pulse ($Y_f^G(I_0)$) as a function of the peak intensity of UV pulses

to the second and higher-lying electronically excited states. Only from these second and higher-lying states $C_2H_5^+$, $C_2H_3^+$, and CH_3^+ can be produced energetically. The relative yield of $Y_f^E(I_0) = 0.74$ is obtained as the sum of the final relative yields of all the electronically excited states of $C_2H_5OH^+$ at $I_0 = 1.5$ TW/cm^2.

2.3.3.3 Intensity Dependence

When the laser peak intensity increases to $I_0 = 17$ TW/cm^2, the spectral features are changed as shown in Fig. 2.7b. The five-photon ATI peaks become more prominent in all the channels. In the spectra for the channels yielding CH_2OH^+ and $C_2H_5^+$, the peaks corresponding to the electronic ground state of $C_2H_5OH^+$ can be identified more clearly than those corresponding to the electronically excited state, indicating the branching ratio of the electronic excitation subsequent to the ionization increases to 0.7 for CH_2OH^+ and 0.8 for $C_2H_5^+$ with respect to the respective product ion yields.

The PEPICO measurements were performed at the five different laser peak intensities in the range between $I_0 = 1.5$ and 17 TW/cm^2. In the same manner as described in Sect. 2.3.3.2 for $I_0 = 1.5$ TW/cm^2, the channel-specific photoelectron spectra were analyzed to obtain $Y_i^G(I_0)$, $Y_i^E(I_0)$, $Y_f^G(I_0)$, and $Y_f^E(I_0)$. It is shown in Fig. 2.8 that, as I_0 increases from 1.5 to 17 TW/cm^2, $Y_i^G(I_0)$ increases from 0.32 to 0.8 while $Y_f^G(I_0)$ decreases from 0.26 to 0.14.

The Keldysh parameter $\gamma = (I_P/2U_P)^{1/2}$ has been used as a criterion of whether the ionization proceeds through tunneling ($\gamma < 1$) or multiphoton absorption ($\gamma > 1$) [9], where I_P and U_P stand for the ionization potential and the ponderomotive energy, respectively. At $I_0 = 1.5$ TW/cm^2, the Keldysh parameter for the ionization to the electronic ground state of $C_2H_5OH^+$ is $\gamma = 15.2$, suggesting that the multiphoton ionization mechanism is dominant.

The observed result that $Y_i^E(I_0) = 0.68$ is larger than $Y_i^G(I_0) = 0.32$ at $I_0 = 1.5$ TW/cm^2 is interpreted by the multiphoton ionization mechanism as described below. The probability of the multiphoton ionization to the jth electronic state of a parent ion is proportional to a product of the N-photon cross section $\sigma_j^{(N)}$ and the Nth power of I_0 as

$$w_j(I_0) \quad \propto \quad \sigma_j^{(N)} I_0^N, \tag{2.1}$$

where $\sigma_j^{(N)}$ is independent of the laser intensity and is determined by the N-photon transition dipole moment between the electronic ground state of a neutral molecule and the ionization continuum correlated with the jth electronic states of a cation. In the ionization of C_2H_5OH by the 400 nm fields, four photons are required for the ionization to both the electronic ground state and the first electronically excited state. Therefore, the nascent relative yields of the electronic ground state ($j = 0$) and the first electronically excited state ($j = 1$) are determined by $\sigma_0^{(4)}$ and $\sigma_1^{(4)}$, respectively. The nascent populations in the second and higher electronically excited states of $C_2H_5OH^+$ at the moment of photoelectron emission was not identified in the present measurement, possibly because one more photon is required for their formation at the photoelectron emission.

As the intensity I_0 increases from 1.5 to 17 TW/cm^2, $Y_i^G(I_0)$ increases from 0.32 to 0.8. Within the multiphoton ionization mechanism whose ionization rate is expressed as (2.1), the electronic state distribution at the moment of the photoelectron emission is determined by $\sigma_j^{(N)}$ while the four photon ionization probabilities of both the electronic ground state and the first electronically excited state are proportional to I_0^4. Accordingly, the observed change in the electronic state distribution as a function of I_0 cannot be interpreted simply by the multiphoton ionization mechanism.

It should be noted that as I_0 increases the ionization threshold increases by the poderomotive energy up-shift, resulting in the channel closing [47]. As shown in Fig. 2.7a, the broad peak corresponding to the first electronically excited state of $C_2H_5OH^+$ appears at 0.5 eV in the photoelectron spectrum for the CH_2OH^+ formation at $I_0 = 1.5$ TW/cm^2, where the ponderomotive energy shift is 0.02 eV. As the laser intensity increases to $I_0 = 17$ TW/cm^2, the ponderomotive energy shift becomes 0.26 eV, which is 0.24 eV larger than that at $I_0 = 1.5$ TW/cm^2. Assuming that (i) the ionization channel corresponding to the peak area in the electron energy range from 0 to 0.24 eV in the spectrum for the CH_2OH^+ formation shown in Fig. 2.7a is closed by the increase in I_0 from 1.5 to 17 TW/cm^2 and (ii) the ionization to the electronic ground state is not closed, the ionization yield to the first electronically excited state is estimated to decrease by only 13 %, which is much smaller than the observed decrease of 88 % in the ratio $Y_i^E(I_0)/Y_i^G(I_0)$. This discrepancy suggests that another mechanism plays a dominant role in increasing $Y_i^G(I_0)$ while decreasing $Y_i^E(I_0)$.

The observed result that $Y_i^G(I_0)$ increases as I_0 increases from $I_0 = 1.5$ to 17 TW/cm^2 can be regarded as a transition to the tunneling ionization that the ionization probability decreases exponentially as a function of the ionization potential [12, 48]. On the other hand, the ionization to the electronic ground state of $C_2H_5OH^+$ is still in the multiphoton ionization regime when judged only by the Keldysh parameter, $\gamma = 4.5$ (>1) at $I = 17$ TW/cm^2. This contradictory situation suggests that the

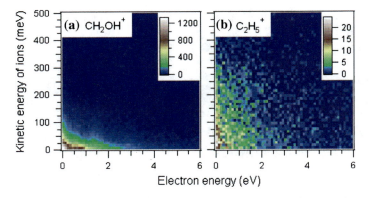

Fig. 2.9 Energy correlation maps of a photoelectron and a fragment ion for the channels yielding **a** CH_2OH^+ and **b** $C_2H_5^+$ when $\lambda = 783$ nm, $I_0 = 9$ TWcm^2, and $\tau = 35$ fs

ionization mechanism may not be discussed simply based on the Keldysh parameter that tends to be overestimated owing to the assumption of the zero-range potential in deriving the tunneling ionization probability [14].

2.4 Correlation Between a Photoelectron and a Fragment Ion

2.4.1 Energy Correlation Mapping

From the 3D momenta of fragment ions as well as photoelectrons observed in our measurement, not only the channel-specific photoelectron spectra but also the energy correlation between a photoelectron and a fragment ion can be obtained as shown in Fig. 2.9 when $\lambda \sim 783$ nm, $I_0 = 9$ TW/cm^2, and $\tau = 35$ fs. From these energy correlation maps, the kinetic energy distributions of the fragment ions are extracted as a function of photoelectron energy E_{elec} [36]. For instance, the kinetic energy distributions of CH_2OH^+ are shown in Fig. 2.10a, b at $E_{elec} = 0.8$ and 1.5 eV, respectively. These distributions $P_{ion}(E_{ion})$ are reproduced well by a Boltzmann-type distribution expressed as

$$P_{ion}(E_{ion}) = A\exp(-E_{ion}/kT), \qquad (2.2)$$

where T represents the temperature of a fragment ion, A an arbitrary coefficient, and k the Boltzmann constant. This Boltzmann-type distribution suggests that the internal energy gained from the laser field is statistically distributed among the vibrational modes of $C_2H_5OH^+$ prior to the dissociation. The least-squares fitting to (2.2) was performed to obtain the temperature T at the E_{elec} values in the range of $E_{elec} = 0 - 6$ eV with an interval of 0.1 eV as plotted in Fig. 2.11a.

Fig. 2.10 Kinetic energy distributions (*diamond*) of CH_2OH^+ extracted from the energy correlation map shown in Fig. 2.9a. The *dotted lines* are results of the least-square fitting to (2.2). The temperatures obtained by the least-square fitting are 284(6) and 465(6) K in **a** $E_{elec} = 0.8$ eV and **b** $E_{elec} = 0.15$ eV, respectively

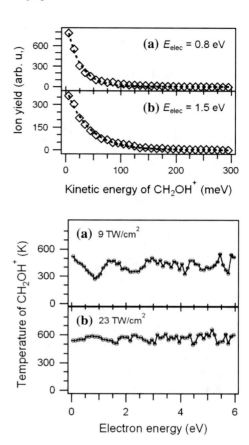

Fig. 2.11 Translational temperature of CH_2OH^+ as a function of E_{elec} when $\lambda \sim 783$ nm, $\tau = 35$ fs, and $I_0 =$ **a** 9 and **b** 23 TW/cm^2

2.4.2 Translational Temperature of Fragment Ions

The temperature of CH_2OH^+ exhibits oscillation as a function of E_{elec} when $\lambda \sim 783$ nm, $\tau = 35$ fs, and $I_0 = 9$ TW/cm^2. As the photoelectron energy increases from $E_{elec} = 0$ eV, the temperature decreases to the minimum value of 267 K at $E_{elec} \sim 0.8$ eV, where the ionization to the electronically excited state of $C_2H_5OH^+$ dominantly proceeds. As E_{elec} increases from $E_{elec} = 0.8$ eV, the temperature of CH_2OH^+ starts to increase and takes the maximum value of 464 K at $E_{elec} \sim 1.5$ eV, which coincides with the energy of the eight-photon ionization of C_2H_5OH to form the electronic ground state of $C_2H_5OH^+$ as seen in the photoelectron spectra in Fig. 2.5a for the channels yielding $C_2H_5OH^+$ and $C_2H_4OH^+$. As E_{elec} increases further from $E_{elec} = 1.5$ eV, the temperature of CH_2OH^+ decreases to the minimum value of 348 K at $E_{elec} \sim 2.3$ eV, and then increases again to the maximum of 498 K at $E_{elec} \sim 3.0$ eV.

The oscillation of the temperature of CH_2OH^+ as a function of E_{elec} suggests that there are two different types of ionization and electronic excitation pathways. When the photoelectron energy in the channel yielding CH_2OH^+ is close to the ATI peak energies in the photoelectron spectra for the channels yielding $C_2H_5OH^+$ and $C_2H_4OH^+$, the ionization to the electronic ground state of $C_2H_5OH^+$ is promoted, and then, the electronic excitation subsequent to the ionization proceeds to yield CH_2OH^+. When the photoelectron energy is apart from the ATI peak energies for the channels yielding $C_2H_5OH^+$ and $C_2H_4OH^+$, the ionization to the electronically excited state of $C_2H_5OH^+$ becomes dominant. It should be noted that these two different types of ionization and electronic excitation pathways have been identified from the oscillatory behavior of the temperature of CH_2OH^+ even though the photoelectron spectrum for the channel yielding CH_2OH^+ exhibits the diffuse spectral feature with no structured feature.

When the electronic ground state of $C_2H_5OH^+$ is formed at the moment of photoelectron emission, the subsequent electronic excitation is required for the CH_2OH^+ formation because CH_2OH^+ cannot be produced energetically from the electronic ground state of $C_2H_5OH^+$ [40]. On the other hand, the relatively low value of the temperature of CH_2OH^+ in the E_{elec} region between the ATI peaks reflects the fragmentation processes to form CH_2OH^+ from the electronically excited state of $C_2H_5OH^+$ prepared at the moment of photoelectron emission. These findings indicate that the internal energy of $C_2H_5OH^+$ gained through the electronic excitation after the photoelectron emission to form $C_2H_5OH^+$ in the electronic ground state is larger than that of the electronically excited $C_2H_5OH^+$ prepared at the moment of photoelectron emission. On the basis of the energy correlation between a photoelectron and a product ion, it has been revealed that $C_2H_5OH^+$ prepared in the lower electronic states at the moment of photoelectron emission is excited into the higher-lying electronic states after the excitation by the laser pulse. It should be emphasized that this excitation dynamics can be extracted through the energy correlation measurement of a photoelectron and a fragment ion.

As I_0 increases to 23 TW/cm^2 while keeping $\tau = 35$ fs, the oscillation amplitude of the temperature of CH_2OH^+ as a function of E_{elec} is reduced as shown in Fig. 2.11b. The energies at which the maximum and minimum temperatures are taken are shifted by around 0.7 eV towards the smaller energy side because of the ponderomotive energy shift of the ionization threshold energies, which can be seen in the photoelectron spectra in Fig. 2.5a, c. It is also indicated that the maximum (and averaged) temperature of CH_2OH^+ increases as I_0 increases. It should be noted that the ATI peaks become more distinct in the photoelectron spectrum for the channel yielding CH_2OH^+ in Fig. 2.5c, suggesting that the electronic ground state of $C_2H_5OH^+$ is dominantly formed at the moment of photoelectron emission. The increase in the maximum temperature of CH_2OH^+ with increasing I_0 means that $C_2H_5OH^+$ gains larger internal energy as I_0 increases.

The above analysis on the channel yielding CH_2OH^+ was similarly performed on the channel yielding $C_2H_5^+$. The kinetic energy distributions of $C_2H_5^+$ as a function of E_{elec} are reproduced by the Boltzmann-type distributions. The temperatures of $C_2H_5^+$ obtained by the least-squares fit are plotted in Fig. 2.12a as a function of E_{elec}

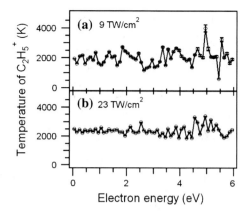

Fig. 2.12 Translational temperature of $C_2H_5^+$ as a function of E_{elec} when $\lambda \sim 783$ nm, $\tau = 35$ fs, and $I_0 =$ **a** 9 and **b** 23 TW/cm^2. The large fluctuation in (**a**) is mainly ascribed to the fact that the number of the accumulated events for the channel yielding $C_2H_5^+$ is not large enough

when $\lambda \sim 783$ nm, $I_0 = 9$ TW/cm^2, and $\tau = 35$ fs. The higher temperature of $C_2H_5^+$ in the range between 1150 and 2700 K than the temperature of CH_2OH^+ in the range between 267 and 520 K can be ascribed to larger excess energy in the channel yielding $C_2H_5^+$ [49]. At $I_0 = 23$ TW/cm^2, as seen in Fig. 2.12b, the temperature of $C_2H_5^+$ exhibits almost constant values as a function of E_{elec}, and an oscillatory behavior cannot be recognized clearly, indicating that the internal state distribution of $C_2H_5OH^+$ prepared at the moment of photoelectron emission is not coupled with the electronic excitation process of $C_2H_5OH^+$ after the photoelectron emission.

Dependences of the translational temperature of CH_2OH^+ and $C_2H_5^+$ on the laser pulse duration were also investigated by chirping positively and negatively while keeping the peak intensity I_0 in the range between 17 and 19 TW/cm^2 [36]. An oscillatory behavior cannot be identified clearly in the temperature of CH_2OH^+ and $C_2H_5^+$ when the pulses are stretched to $\tau = 200$ and 800 fs. This result suggests that, as the pulse duration τ increases, the electronic excitation of $C_2H_5OH^+$ proceeds to a large extent within a laser pulse, and the population prepared in the electronically excited state at the moment of photoelectron emission is also suppressed. Consistently, as τ increases, the distinct ATI peaks become more prominent in the photoelectron spectra for the channels yielding CH_2OH^+ and $C_2H_5^+$ similarly to those in the photoelectron spectra for the channels yielding $C_2H_5OH^+$ and $C_2H_4OH^+$ as shown in Fig. 2.6. These photoelectron spectra show that the branching ratio of the formation of the electronically excited state at the moment of the photoelectron emission decreases as τ increases.

2.5 Summary

We have reviewed our recent studies on the multichannel dissociative ionization of ethanol induced by intense laser fields by the photoelectron-photoion coincidence momentum imaging. From the channel-specific photoelectron spectra, it was found that there are two types of ionization and dissociation pathways in the dissociative ionization dynamics of C_2H_5OH. One pathway is the ionization to form the electronic ground state of $C_2H_5OH^+$ followed by the electronic excitation leading to the dissociation. The other is the ionization into the electronically excited state of $C_2H_5OH^+$, from which the dissociation is energetically allowed. The energy correlation between a photoelectron and a fragment ion indicates how $C_2H_5OH^+$ gains the internal energy from the laser field. Furthermore, branching into the ionization and subsequent electronic excitation pathways was found to depend sensitively on laser intensity I_0, pulse duration τ, and wavelength λ. Interestingly, the population in the electronic ground state of $C_2H_5OH^+$ at the photoelectron emission increases as I_0 increases irrespective of the wavelength.

Acknowledgments Our study described in this chapter was supported in part by the JAEA special research project "Reaction control in intense laser fields", by MEXT/JSPS KAKENHI (Grant numbers 14077205, 17750004, 19685003, and 23350013) and by the Matsuo Foundation.

References

1. L.J. Frasinski et al., Femtosecond dynamics of multielectron dissociative ionization by use of picosecond laser. Phys. Rev. Lett. **58**, 2424 (1987)
2. C. Cornaggia et al., Multielectron dissociative ionization of diatomic molecules in an intense femtosecond laser field. Phys. Rev. A **44**, 4499 (1991)
3. S.L. Chin, Multiphoton ionization of molecules. Phys. Rev. A **4**, 992 (1971)
4. A. Talebpour, A.D. Bandrauk, J. Yang, S.L. Chin, Multiphoton ionization of inner-valence electrons and fragmentation of ethylene in an intense Ti:sapphire laser pulse. Chem. Phys. Lett. **313**, 789 (1999)
5. M. Lezius, V. Blanchet, M.Y. Ivanov, A. Stolow, Polyatomic molecules in strong laser fields: nonadiabatic multielectron dynamics. J. Chem. Phys. **117**, 1575 (2002)
6. A.N. Markevitch et al., Sequential nonadiabatic excitation of large molecules and ions driven by strong laser fields. Phys. Rev. A **69**, 013401 (2004)
7. H. Kono et al., Quantum mechanical study of electronic and nuclear dynamics of molecules in intense laser fields. Chem. Phys. **304**, 203 (2004)
8. H. Kono et al., Theoretical investigations of the electronic and nuclear dynamics of molecules in intense laser fields: quantum mechanical wave packet approaches. Bull. Chem. Soc. Jpn. **79**, 196 (2006)
9. L.V. Keldysh, Ionization in the field of a strong electromagnetic wave. Zh. Eksp. Teor. Fiz. **47**, 1945 (1964). [Soviet Physics JETP **20**, 1307–1314 (1965)]
10. F.H.M. Faisal, Multiphoton ionization of hydrogenic atoms and the breakdown of perturbative calculations at high intensities. J. Phys. B **6**, 584 (1973)
11. H.R. Reiss, Effect of an intense electromagnetic field on a weakly bound system. Phys. Rev. A **22**, 1786 (1980)
12. A. Becker, F.H.M. Faisal, Intense-field many-body S -matrix theory. J. Phys. B **38**, R1 (2005)
13. E. Mevel et al., Atoms in strong optical fields: evolution from multiphoton to tunnel ionization. Phys. Rev. Lett. **70**, 406 (1993)
14. M.J. DeWitt, R.J. Levis, Observing the transition from a multiphoton-dominated to a field-mediated ionization process for polyatomic molecules in intense laser fields. Phys. Rev. Lett. **81**, 5101 (1998)
15. B.K. McFarland, J.P. Farrell, P.H. Bucksbaum, M. Guhr, High harmonic generation from multiple orbitals in N_2. Science **322**, 1232 (2008)

2 Electron and Ion Coincidence Momentum Imaging

16. O. Smirnova et al., High harmonic interferometry of multi-electron dynamics in molecules. Nature **460**, 972 (2009)
17. S. Haessler et al., Attosecond imaging of molecular electronic wavepackets. Nat. Phys. **6**, 200 (2010)
18. L. Fang, G.N. Gibson, Investigating excited electronic states of I_0^+ and I_2^{2+} produced by strong-field ionization using vibrational wave packets. Phys. Rev. A **75**, 063410 (2007)
19. D. Geissler et al., Creation of multihole molecular wave packets via strong-field ionization. Phys. Rev. A **82**, 011402(R) (2010)
20. M. Kotur et al., Neutral-ionic state correlations in strong-field molecular ionization. Phys. Rev. Lett. **109**, 203007 (2012)
21. J. Mikosch et al., Channel- and angle-resolved above threshold ionization in the molecular frame. Phys. Rev. Lett. **110**, 023004 (2013)
22. M. Spanner, S. Patchkovskii, One-electron ionization of multielectron systems in strong non-resonant laser fields. Phys. Rev. A **80**, 063411 (2009)
23. A.E. Boguslavskiy et al., The multielectron ionization dynamics underlying attosecond strong-field spectroscopies. Science **335**, 1336 (2012)
24. J.H.D. Eland, Angular distributions, energy disposal, and branching studied by photoelectron-photoion coincidence spectroscopy: O_2^+, NO^+, ICl^+, IBr^+, and I_2^+ fragmentation. J. Chem. Phys. **70**, 2926 (1979)
25. E. Shigemasa, J. Adachi, M. Oura, A. Yagishita, Angular distributions of $1s\sigma$ photoelectrons from fixed-in-space N_2 molecules. Phys. Rev. Lett. **74**, 359 (1995)
26. K. Hosaka et al., Coincidence velocity imaging apparatus for study of angular correlations between photoelectrons and photofragments. Jpn. J. Appl. Phys. **45**, 1841 (2006)
27. T. Osipov et al., Photoelectron-photoion momentum spectroscopy as a clock for chemical rearrangements: isomerization of the Di-Cation of Acetylene to the Vinylidene configuration. Phys. Rev. Lett. **90**, 233002 (2003)
28. J.A. Davies, J.E. LeClaire, R.E. Continetti, C.C. Hayden, Femtosecond time-resolved photoelectron-photoion coincidence imaging studies of dissociation dynamics. J. Chem. Phys. **111**, 1 (1999)
29. J.A. Davies, R.E. Continetti, D.W. Chandler, C.C. Hayden, Femtosecond time-resolved photoelectron angular distributions probed during photodissociation of NO_2. Phys. Rev. Lett. **84**, 5983 (2000)
30. H. Rottke et al., Coincident fragment detection in strong field photoionization and dissociation of H_2. Phys. Rev. Lett. **89**, 013001 (2002)
31. T. Hatamoto et al., High-resolution electron-ion coincidence spectroscopy of ethanol in intense laser fields. Phys. Rev. A **75**, 061402(R) (2007)
32. A. Matsuda, M. Fushitani, A. Hishikawa, Electron-ion coincidence momentum imaging of molecular dissociative ionization in intense laser fields: application to CS_2. J. Electron Spectrosc. Relat. Phenom. **169**, 97 (2009)
33. K. Hosaka et al., Photoelectron-photoion coincidence momentum imaging for dissociative ionization of ethanol in intense laser fields. Chem. Phys. Lett. **475**, 19 (2009)
34. H. Akagi et al., Laser tunnel ionization from multiple orbitals in HCl. Science **325**, 1364 (2009)
35. T. Ikuta et al., Separation of ionization and subsequent electronic excitation for formation of electronically excited ethanol cation in intense laser fields. J. Phys. B **44**, 191002 (2011)
36. K. Hosaka, A. Yokoyama, K. Yamanouchi, R. Itakura, Correlation between a photoelectron and a fragment ion in dissociative ionization of ethanol in intense near-infrared laser fields. J. Chem. Phys. **138**, 204301 (2013)
37. M. Lebech, J.C. Houver, D. Dowek, Ion-electron velocity vector correlations in dissociative photoionization of simple molecules using electrostatic lenses. Rev. Sci. Instrum. **73**, 1866 (2002)
38. R. Trebino et al., Measuring ultrashort laser pulses in the time-frequency domain using frequency-resolved optical gating. Rev. Sci. Instrum. **68**, 3277 (1997)
39. S. Katsumata, T. Iwai, K. Kimura, Photoelectron spectra and sum rule consideration. Higher alkyl amines and alchools. Bull. Chem. Soc. Jpn. **46**, 3391 (1973)

40. Y. Niwa, T. Nishimura, T. Tsuchiya, Ionic dissociation of ethanol studied by photoelectron-photoion coincidence spectroscopy. Int. J. Mass Spectrom. Ion Phys. **42**, 91 (1982)
41. K. Kimura et al., *Handbook of HeI Photoelectron Spectra of Fundamental Organic Molecules* (Japan Scientific Societies Press, Tokyo, 1981)
42. H. Niikura et al., Sub-laser-cycle electron pulses for probing molecular dynamics. Nature **417**, 917 (2002)
43. X. Xie et al., Attosecond-recollision-controlled selective fragmentation of polyatomic molecules. Phys. Rev. Lett. **109**, 243001 (2012)
44. I.V. Litvinyuk et al., Shakeup excitation during Optical Tunnel Ionization. Phys. Rev. Lett. **94**, 033003 (2005)
45. R.R. Freeman et al., Above-threshold ionization with subpicosecond laser pulses. Phys. Rev. Lett. **59**, 1092 (1987)
46. M.B. Robin, *Higher Excited States of Polyatomic Molecules*, vol. 1 (Academic Press, New York, 1974)
47. A.M. Muller et al., Photoionization and photofragmentation of gaseous toluene using 80 fs, 800 nm laser pulses. J. Chem. Phys. **112**, 9289 (2000)
48. M.V. Ammosov, N.B. Delone, V.P. Krainov, Tunnel ionization of complex atoms and of atomic ions in an alternating electromagnetic field. Zh. Eksp. Teor. Fiz. **91**, 2008 (1986). [Sov. Phys. JETP **64**, 1191–1194 (1986)]
49. H.-F. Lu et al., Fragmentations of singly charged ethanol cation: an ab initio/RRKM study. J. Mol. Struct. Theochem. **761**, 159 (2006)

Chapter 3
Exploring and Controlling Fragmentation of Polyatomic Molecules with Few-Cycle Laser Pulses

Markus Kitzler, Xinhua Xie and Andrius Baltuška

Abstract The removal of electrons from polyatomic molecules by ionization with intense, ultrashort laser pulses may trigger complex restructuring and fragmentation dynamics. Depending on the valence-shell from which the electrons are removed the molecular ion might be put into a certain dissociative or binding state by the ionization process. With control over the ionization process it might thus be possible to gain control over the subsequent restructuring and fragmentation process on a purely electronic level. Here we introduce two conceptually similar schemes that allow controlling the outcome of molecular restructuring and fragmentation processes in polyatomic molecules on sub-femtosecond time-scales. The first one involves recollision double ionization in few-cycle laser fields with a known carrier-envelope phase (CEP). We demonstrate experimentally CEP-control over various fragmentation reactions of a series of polyatomic molecules (acetylene, ethylene, 1,3-butadiene). As the recollision energy for a given intensity sensitively depends on the CEP, tuning of the CEP allows controlling the removal of inner-valence electrons and the controlled population of dissociative excited states. The second control scheme uses the strong preponderance of ionization from specific molecular orbitals to the alignment of the molecular axis with respect to the laser polarization direction for determining which valence level the electrons are removed from. We demonstrate experimental control over different two-body fragmentation and dissociation pathways from the cation and the dication of the acetylene molecule using the field-free alignment angle as a control knob. Finally, we turn from the demonstration of control schemes working at sub-femtosecond time-scales, for which the nuclear dynamics following the ionization are not essential, to the investigation of the coupled nuclear-electronic dynamics that in general take place for longer pulse durations. We explore experimentally the mechanism behind the surprisingly high charge states recently observed in the ionization of hydrocarbon molecules that have been explained by a

---

M. Kitzler (✉) · X. Xie · A. Baltuška
Photonics Institute, Vienna University of Technology, Gusshausstrasse 27/387,
1040 Vienna, Austria
e-mail: markus.kitzler@tuwien.ac.at

K. Yamanouchi et al. (eds.), *Progress in Ultrafast Intense Laser Science XI*,
Springer Series in Chemical Physics 109, DOI: 10.1007/978-3-319-06731-5_3,
© Springer International Publishing Switzerland 2015

multi-bond version of the well known enhanced-ionization (EI) mechanism taking place in parallel at many C–H bonds. Our experimental results are in agreement with the proposed multi-bond version of EI.

3.1 Introduction

Creation of electronic wavepackets of sub-femtosecond duration, a pre-requisite for tracing and imaging structural dynamics on the attosecond time-scale, is facilitated by the process of strong-field ionization that takes place within a short time interval at the peaks of an intense laser half-cycle. In contrast to atoms, for which this process is now relatively well understood, there are a number of characteristics and accompanying processes that render ionization of molecules much more complicated and make it an active field of research.

A striking difference is that the ionization rate in molecules sensitively depends on the orientation of the molecule with respect to the laser polarization axis and on the electronic structure [1–8]. In addition, the energy levels in molecules can be relatively closely spaced such that, within a Hartree-Fock picture, ionization may also take place from lower-lying valence orbitals with a pronounced probability, e.g. [9–14], especially when ionization from the highest lying states is suppressed due to unfavorable structure for a given molecular alignment or orientation.

Furthermore, the laser field may also drive non-adiabatic electronic excitations to higher (ionic) energy levels [15–18] from which ionization can proceed more easily during subsequent field cycles. Excitation of a molecule or molecular ion can also be induced by electron recollision, a process originally discovered and studied for atoms [19–21]. Often, excited states in a molecule or molecular ion are dissociative. Therefore, excitations taking place during ionization may lead to fragmentation of the molecule [14, 22]. For example, excitations due to ionization from a lower lying orbital induced by electron recollision can be used to control fragmentation reactions of polyatomic molecules on the laser-sub-cycle time scale [23].

Finally, the impinging laser pulse may also trigger nuclear dynamics and molecular rearrangement processes which might lead to a strongly varying ionization rate as the nuclear motion proceeds. One of the most striking manifestations of this effect is enhanced ionization (EI) [24, 25], where at a critical internuclear distance, R_c, the ionization rate is strongly enhanced as compared to the equilibrium internuclear distance or the dissociative limit. By the development of the concept of enhanced ionization it became possible to explain the energy of ionic fragments observed in experiments on laser ionization of molecules, e.g. [26–30], which were found to be consistently below the Coulomb explosion limit. Since then EI has been investigated experimentally many times in diatomics, e.g. [31–33], and recently also for the triatomic molecule CO_2 [34].

The goal of this chapter is to explore the dependence of the restructuring and subsequent fragmentation dynamics of polyatomic molecules on the sub-cycle-triggered removal of specific valence electrons by strong-field ionization with few-cycle laser

3 Exploring and Controlling Fragmentation

pulses. In our experiments we study polyatomic molecules of different size ranging from four to ten atoms. Specifically, because of their very rich and fast restructuring and fragmentation dynamics due to the light and quickly responding hydrogen atoms, we investigate the hydrocarbon molecules acetylene, C_2H_2, ethylene, C_2H_4, and 1,3-butadiene, C_4H_6.

Starting by reviewing our recent result [23] that the controlled removal of lower-valence electrons by recollision-ionization [21, 35–38] with few-cycle laser pulses with a known carrier-envelope phase (CEP) allows determining the outcome of various fragmentation reactions in polyatomic molecules, we then discuss whether there is a possibility to selectively remove electrons from *specific* (inner-)valence orbitals. This could potentially allow the design of schemes for controlling molecular fragmentation reactions via ionization to excited dissociative molecular states [23] that are capable of not only enhancing or suppressing molecular fragmentation reactions collectively for several pathways, but in addition also allow to distinguish between different fragmentation pathways. We will show that this might be possible by exploiting the strong preponderance of ionization from specific molecular orbitals to the alignment of the molecular axis with respect to the laser polarization direction [2, 3].

Finally, we will move our attention from molecular dissociation and fragmentation reactions starting from the singly or doubly charged molecular ions to extreme cases of molecular fragmentation starting from surprisingly high charge states recently observed in [39]. In that paper it was shown that with 27 fs laser pulses at 790 nm very high charge states (e.g. $\geq +13$ for C_4H_6) could be reached already at moderate intensities of a few 10^{14} W/cm^2. These high charge states result in complete fragmentation of the molecular ions and the emission of protons with considerable kinetic energy. Ionization to such high charge states obviously necessitates the involvement of electrons in inner-valence orbitals in the underlying ionization process. We describe experiments aimed at investigating the coupled electronic and nuclear dynamics behind the ionization process that leads to the highly charged hydrocarbon molecules.

3.2 Controlling Fragmentation Reactions with the Shape of Intense Few-Cycle Laser Pulses

Electronic and nuclear dynamics triggered by the removal of electrons in a strong laser field may result in the fragmentation of the molecule. Because the electron release by field ionization is confined to a sub-cycle interval at the peak of an intense optical half-cycle, the outcome of the fragmentation reaction is to a large extent predetermined on a sub-cycle time scale. Consequently, as has been shown in the case of diatomic molecules—D_2 [40–42], H_2 [43, 44], HD [41], DCl [45] and CO [46, 47]—dissociative ionization strongly depends on the CEP of a few-cycle ionizing laser pulse. The CEP-dependence in the case of the lightest molecules,

D_2/H_2, has been linked to a field-driven population transfer between the binding ground state and the dissociative excited state.

In the following we review our recent results [23] on the first CEP control of fragmentation pathways of polyatomic molecules (acetylene, ethylene, 1,3-butadiene). Clearly, because of the increased number of participating nuclei and a vastly more complex valence electron dynamics and the structure of energy surfaces, it was all but clear that the CEP-dependence of fragmentation pathways survives in polyatomic molecules. Yet, our experiments revealed a prominent role of the CEP for all three polyatomic species under examination. We will show below that the key to determining the outcome of a fragmentation reaction in polyatomic molecules is a precise control over the process that triggers the fragmentation, i.e. the ionization step. In particular we will demonstrate that removal of electrons from lower lying orbitals, which puts the resulting molecular ion into an electronically excited state, can be used to control fragmentation reactions of polyatomic molecules. As the selective removal of electrons from specific orbitals can be initiated on the laser-sub-cycle time scale, it can constitute a universal attosecond mechanism of quasi-single-cycle fragmentation control. In this section we will focus on the controlled removal of electrons by electron recollision [48]. In Sect. 3.3 we will show that the concept of controlling molecular fragmentation reactions by selectively removing electrons from certain (inner-valence) orbitals can be refined if the molecules are aligned with respect to the laser polarization direction.

3.2.1 Experiment

In our experiments, we focused few-cycle laser pulses with sub-5 fs duration in an ultra-high vacuum chamber into a supersonic molecular gas jet of randomly aligned acetylene (C_2H_2), ethylene (C_2H_4), or 1,3-butadiene (C_4H_6) molecules. The few-cycle pulses were generated by spectral broadening of ≈ 25 fs (FWHM) laser pulses from a Titanium-Sapphire laser amplifier system, running at 5 kHz repetition rate, in a 1 m long hollow-core glass capillary, filled with neon, and subsequent recompression by several bounces from chirped mirrors. The electric field of the laser pulses can be written as $E(t) = \mathcal{E}(t)\cos(\omega t + \varphi)$, where $\mathcal{E}(t)$ is the pulse envelope and ω and φ are the light frequency and the carrier-envelope phase (CEP), respectively. The few-cycle pulses, linearly polarized along the z-direction of our lab coordinate system, with their spectra centered around 750 nm, are directed into the ultra-high vacuum chamber (background pressure 1.3×10^{-10} mbar), where they are focused into the supersonic molecular gas jet by a spherical mirror with a focal length of 150 mm, see Fig. 3.1 for an illustration. The gas jet, propagating along the x-direction, was created by expanding the molecular gas samples, with a backing pressure of around 0.5 bar (depending on the molecular species), through a nozzle of 10 μm in diameter into a vacuum chamber at $\sim 4 \times 10^{-5}$ mbar operating pressure and subsequent two-stage skimming and differential pumping. To avoid smearing of the CEP due to the Gouy phase shift when the laser pulses propagate along the y-direction through the gas

3 Exploring and Controlling Fragmentation

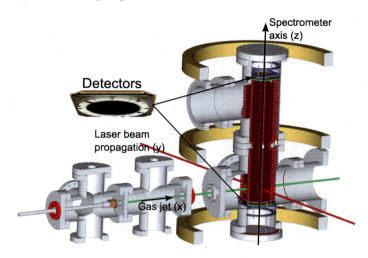

Fig. 3.1 Detection apparatus used for the measurement of coincidence ion momentum spectra

jet [49], which is ≈170 μm in diameter, the jet is cut in width by an adjustable slit before it reaches the interaction region. This reduces the width of the jet to typically 30 μm. The laser focus is placed about half the Rayleigh length before the center of the narrow jet. The Rayleigh length is estimated as 150 μm.

Cold Target Recoil Ion Momentum Spectroscopy (COLTRIMS) [50, 51] was used to measure the three-dimensional momentum vector of fragment ions emerging from the interaction of a single molecule with a few-cycle laser pulse. The ions created by the laser pulses were guided over a distance of 5.7 cm by a weak homogeneous electric field of 16 V/cm, applied along the axis of the spectrometer, z, onto a multi-hit capable RoentDek DLD 80 detector (80 mm in diameter), equipped with position sensitive delay line anodes. The ion count rate was kept at approximately 0.4 per laser shot in order to establish coincidence conditions, which ensure that all observed processes take place within a single molecule. From the measured time-of-flight and position of each detected ion we calculated its three-dimensional momentum vector in the lab frame [50, 51]. The peak intensity of the laser pulses on target was determined from separate calibration measurements using single ionization of argon atoms in circularly polarized light [52, 53].

As in our experiments we measure the ion momenta of fragments from only one molecule created by ionization in a single laser pulse, it becomes possible to perform single-shot CEP-tagging [59]: We measure the duration and the CEP, φ, of each few-cycle laser pulse on a single-shot basis by exploiting the asymmetry of photoelectron spectra created by above-threshold ionization (ATI), emitted to the left and right along the laser polarization direction, using a stereo-ATI phase-meter [60–62]. In the offline data analysis each fragmentation event is then tagged with the corresponding CEP-value determined from the photoelectron spectra [61, 62]. From

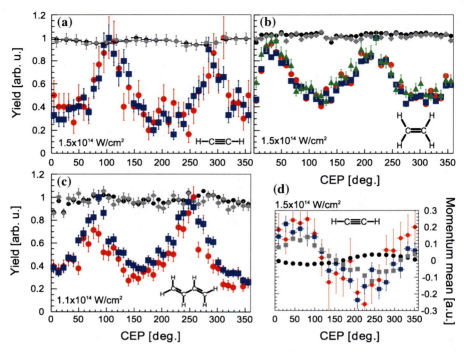

Fig. 3.2 a–c Measured fragmentation and ionization yields, normalized to 1 at their respective maxima, as a function of CEP for different fragmentation channels of acetylene (**a**), ethylene (**b**), and 1,3-butadiene (**c**), measured at the laser intensities indicated in the panels. The various fragmentation reactions are $C_2H_2^{2+} \rightarrow CH^+ + CH^+$ (*dots*) and $C_2H_2^{2+} \rightarrow H^+ + C_2H^+$ (*squares*) for acetylene (**a**); $C_2H_4^{2+} \rightarrow CH_2^+ + CH_2^+$ (*dots*), $C_2H_4^{2+} \rightarrow H^+ + C_2H_3^+$ (*squares*), and $C_2H_4^{2+} \rightarrow C_2H_2^+ + H_2^+$ (*triangles*) for ethylene (**b**); $C_4H_6^{2+} \rightarrow C_2H_3^+ + C_2H_3^+$ (*dots*) and $C_4H_6^{2+} \rightarrow CH_3^+ + C_3H_3^+$ (*squares*) for 1,3-butadiene (**c**). The normalized ionization yields of the singly and doubly charged molecular ions are approximately independent of the CEP and shown by *dots* and *squares*, respectively, around the value of 1 in each panel. **d** Mean p_z value (integrated over p_x and p_y) of the singly (*dots*) and doubly (*squares*) charged molecular ions of acetylene as a function of CEP, in comparison with the p_z-sum of the two fragments from the two fragmentation channels shown in the same point style as in panel (**a**)

the stereo-ATI data we estimate the duration of our laser pulses as 4.5 fs full width at half maximum (FWHM).

As CEP-measurement and coincidence measurement of fragment ions are performed in two separate apparatus, the laser beam, split into two arms, passes through different amounts of glass and air. Thus, the CEP-value determined from the stereo-ATI phase-meter, φ, is offset by a constant value, φ_0, from the one of the pulses in the COLTRIMS device, φ_{CE}. We arbitrarily calibrated the CEP in the COLTRIMS device such that the mean momentum value of the singly charged molecular ion, which due to momentum conservation is the inverse of the electron momentum, peaks at $\varphi_{CE} = 90°/270°$ and is zero at $\varphi_{CE} = 0°/180°$, see Fig. 3.2d. This CEP-dependence

3 Exploring and Controlling Fragmentation 49

is predicted by the simple man's model (SMM) of strong field physics [63], which neglects the influence of the long-range Coulomb potential onto the departing electrons. Due to the Coulomb influence, depending on the atomic or molecular species and on the laser parameters, the CEP-values at which the maxima of the mean momentum are observed, do not necessarily coincide with the predictions of the SMM [64, 65]. As the Coulomb influence and laser parameters are different for the separate measurements shown below, φ_0 might also be different for each of them. We would like to emphasize, though, that all explanations and conclusions given below are unaffected by a constant CEP-offset, $\varphi_0 = \varphi - \varphi_{CE}$.

3.2.2 Dependence of Fragmentation Yield on CEP

The main experimental results are summarized in Fig. 3.2. Panels (a)–(c) show the ion yields of different two-body fragmentation channels from the doubly charged ions of acetylene, ethylene and 1,3-butadiene as a function of the CEP, φ_{CE}. A certain two-body fragmentation pathway is uniquely identified from the measured data using momentum conservation conditions in all three spatial dimensions [51, 66, 67]. From all detected ions per laser pulse we select only those sets of two fragment ions, for which the absolute value of their momentum sum is smaller than 3 a.u. along the y and z directions, and smaller than 5 a.u. along the x direction. By that we identified two different fragmentation pathways for acetylene and 1,3-butadiene, and three different ones for ethylene. The fragmentation yields in Fig. 3.2 are the integrals over the three-dimensional momentum spectra of a certain fragmentation channel. While the mean momenta of the fragments show only a slight dependence on the CEP, the yields of all identified fragmentation channels in Fig. 3.2 exhibit an extraordinarily strong dependence on the CEP with the strongest modulation observed for acetylene (close to 80 %). The yields of the singly and doubly charged molecular ions, also shown in Fig. 3.2, are widely independent of the CEP. The fragmentation yield of each of the identified channels, in contrast, is strongly modulated by the CEP.

3.2.3 Discussion of the Underlying Control Mechanism

In the following we will discuss which light-field dependent mechanism is responsible for the observed strong CEP-dependence of the fragmentation yield. We will discuss this question for the example of acetylene, for which the energy level diagram for the C–C stretch mode is shown in Fig. 3.3a. The energy level diagram shows that the dicationic ground state is bound. The observed fragmentation reactions, thus, have to take place on one (or more) excited states. The first dissociative state lies about 5 eV above the dicationic ground state and can lead to both fragmentation channels observed for acetylene. Hence, to observe the strong modulation of the

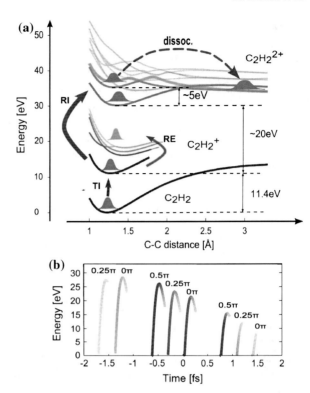

Fig. 3.3 a Electronic energy levels of the neutral, cation and dication of acetylene as a function of C–C distance, calculated by ab initio quantum-chemical methods [54–58]. TI, RI, and RE denote tunnel ionization, recollision ionization and recollision excitation, respectively. **b** Electron recollision energy over the time of ionization calculated for a 4.5 fs (FWHM) long laser pulse with an intensity of 1.5×10^{14} W/cm^2 for the three indicated values of the CEP. The contribution of each recollision event as determined by the ionization probability is indicated by the gray scale intensity

fragmentation yield shown in Fig. 3.2, the population of the involved excited state(s) has to be accomplished by a strongly CEP-dependent mechanism.

As the probability for field ionization depends exponentially on the field strength and, therewith, is sensitive to the CEP of few-cycle pulses, simple sequential double ionization could be a candidate for this mechanism. Indeed, the fragmentation channels observed here have also been observed in other experiments with similarly low laser pulse intensities, but longer pulse durations ($\approx 25 - 50$ fs), e.g. for 1,3-butadiene [67]. In the present experiments, owing to the short duration and low intensity of the laser pulses, and supported by previous measurements with longer pulses [68], we expect double ionization to occur predominantly via electron recollision. Indeed, Fig. 3.2d shows that (i) the measured momentum of $C_2H_2^{2+}$ along the laser polarization direction is much larger than the one of $C_2H_2^+$, and that (ii) the two momenta point into opposite directions for almost all CEP-values. Such CEP-dependence has been shown to be caused by double ionization via electron recollision [38]. Furthermore, as the sum momentum of the two emitted fragments shows the same dependence on the CEP as $C_2H_2^{2+}$ [see Fig. 3.2d], the fragmentation reactions are also initiated by recollision ionization (RI). We thus conclude that the observed CEP-dependence of the fragmentation yields is most likely due to a strong CEP-dependence of RI into the first dissociative states.

3.2.4 Recollision Ionization from Lower-Valence Orbitals

In order to elucidate this point we plot in Fig. 3.3b the electron recollision energy in the few-cycle pulses used in our experiments on acetylene, for different CEP-values as a function of birth time, calculated using the simple man's model [63, 69]. For this intensity (1.5×10^{14} W/cm^2), the strongly CEP-dependent maximum recollision energy of the dominant recollision event is only 22 eV for $\varphi_{CE} = 0$, while for $\varphi_{CE} = \pi/2$ roughly 26 eV are reached. About 20 eV are necessary to reach the ground state of the acetylene dication, see Fig. 3.3a. Additional $\sim$5 eV are necessary to reach the first dissociative state that leads to the experimentally observed correlated fragments CH$^+$/CH$^+$ and C$_2$H$^+$/H$^+$. Thus, the probability for reaching the dissociative state is highest for CEP-values around $\pi/2$, and small for CE-phases around 0. We therefore propose the following scenario [see Fig. 3.3a] for explaining the pronounced fragmentation yield modulations observed in the experiment: An electronic wavepacket set free by field ionization dominantly from the HOMO of the neutral molecule, is driven back by the laser field and doubly ionizes the molecule by electron impact ionization. For CEP-values that result in low recollision energy, the second electron will be dominantly removed from the HOMO, leading to ionization to the non-dissociative dication's ground state. For CEP-values resulting in high recollision energy, the electron can also be removed from a lower lying orbital, which puts the molecular ion into a dissociative excited electronic state. For example, the excited state associated with the fragmentation products CH$^+$/CH$^+$ that we discussed above, is reached by removal of an electron from the HOMO-1 orbital. Because the fragmentation reactions that we investigated in the present experiments take place on electronically excited state surfaces, whose populations are controlled by the CEP, we observe the same CEP dependence for each fragmentation channel of a certain molecule.

3.2.5 Experimental Test of the Recollision Ionization Mechanism

We can test the proposed scenario experimentally simply by increasing the laser peak intensity I, since the recollision energy scales linearly with intensity as $E_{rec} \propto I/4\omega^2$ [63, 69]. If the recollision energy is high enough for all CEP-values to reach the first dissociative state, the fragmentation yield is expected to be independent of the CEP. Figure 3.4a shows the measured yield as a function of CEP of the same fragmentation channels as in Fig. 3.2b, but with a twice higher laser intensity leading to twice the maximum recollision energy. As expected, the fragmentation yields now show no discernible modulation with the CEP. Thus, by precise adjustment of the laser intensity, such that the energy threshold to the dissociative states is overcome only for certain CEP-values, light-field control of the fragmentation of polyatomic molecules becomes possible. This is demonstrated in Fig. 3.4b, for the example of the fragmentation channel C$_4$H$_6^{2+}$ $\rightarrow$ CH$_3^+$ + C$_3$H$_3^+$, where the laser intensity has been

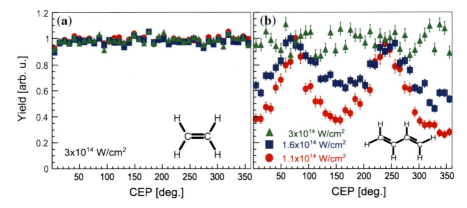

Fig. 3.4 a Fragmentation yields over CEP of the same channels as in Fig. 3.2b (same point style applies), but measured for a slightly higher intensity (as indicated). **b** Measured intensity dependence of the yield of the fragmentation channel $C_4H_6^{2+} \rightarrow CH_3^+ + C_3H_3^+$ over CEP. The intensities are indicated in the figure

fine-tuned in three steps over the dissociation energy threshold, resulting first in a decreased modulation depth of the fragmentation yield, and for even higher intensity in its complete disappearance.

3.2.6 Discussion and Outlook

Electron recollision can not only promote the second electron directly into the continuum, but can also excite the molecular cation, which may then be further ionized by the laser field [21, 70, 71]. We anticipate that this mechanism, dubbed recollision-induced excitation (RE) and subsequent field ionization (RESI), can, in principle, also be used for CEP-control of polyatomic molecular fragmentation with few-cycle pulses even at peak intensities that prohibit direct population of the first excited dicationic state. In this case, however, as many of the closely spaced higher excited states in the cation can be populated by RE, the CEP-selectivity becomes dominantly dependent on the field-sensitive ionization rate during the field ionization step. As a result, the yields of different fragmentation channels as well as the yield of the dication are likely to show different CEP-modulation depths and possibly different CEP-dependence. This is not observed in our experiments. In addition, at the intensity of our experiments and the channels observed, RESI should be considerably less probable than direct RI to the first dissociative excited state. To see this we have estimated the combined probability for recollision excitation to several excited states in the cation [see Fig. 3.3a] and subsequent field ionization, with the latter step modelled using tunneling theory [72]. We found that for the low intensities used in our experiments the probability for this double ionization mechanism is considerably lower than

3 Exploring and Controlling Fragmentation

for direct recollision double ionization. Furthermore, a possible CEP-dependence introduced during the field ionization step should appear also in the yield of the dication. Our experiments, however, do not show any such dependence (see Fig. 3.2). We therefore conclude that the contributions to the observed CEP-modulation of the fragmentation yield due to RESI, while probably important for lower intensities, are insignificant for the here demonstrated higher intensity case that permits direct RI from lower lying orbitals.

Fragmentation reactions of polyatomic molecules are essential building blocks of Chemistry. Equally important, however, are preceding isomerization reactions. In our experiment we observe, for example, two channels that involve molecular restructuring prior to the breakage of a molecular bond: formation of H_2^+ for ethylene and a proton migration reaction for 1,3-butadiene [see Fig. 3.2]. These results demonstrate that RI with intensity-tuned few-cycle laser pulses can be used as a very efficient tool not only to initiate (or suppress) fragmentation but also isomerization reactions in polyatomic molecules with high sensitivity.

3.3 Controlling Fragmentation Reactions by Selective Ionization Into Dissociative Excited States

In Sect. 3.2 we have described that few-cycle pulses with a known CEP can be used to implement an efficient and selective, yet straightforward method of predetermining fragmentation and isomerization reactions in polyatomic molecules on sub-femtosecond time-scales. That method consists in the selective removal of specific (inner-valence shell) electrons, which puts the molecule into a dedicated (dissociative) electronic state from which the molecule then fragments along a desired pathway. Recollision ionization with a precisely tuned recollision energy [23] is, however, not the only method for the selective removal of certain electrons. In this section we will show that the strong preponderance of ionization from specific molecular orbitals to the alignment of the molecular axis with respect to the laser polarization direction [2, 3] allows implementing a very similar type of fragmentation control. We demonstrate control over different two-body fragmentation and dissociation pathways from the cation and the dication of the acetylene molecule (C_2H_2) using the field-free alignment angle as a control knob. The basic principle of the here introduced control method relies on the fact that in many cases a molecular fragmentation reaction can be associated with nuclear dynamics taking place on one (or few) excited (dissociative) electronic surfaces. Population of a specific excited electronic surface can be achieved by the controlled selective removal of electrons from certain inner-valence electronic orbitals. Exploiting the strong angular dependence of the ionization probability from a specific molecular orbital we succeeded in the preferential selective population of certain dissociative excited electronic states, and therewith in the controlled dominant fragmentation along a certain fragmentation pathway associated with this electronic state.

3.3.1 Experiment

The experiments were performed in the ultra-high vacuum chamber of the coincidence detection setup described above, see Fig. 3.1. Initially randomly aligned C_2H_2 molecules supplied by a supersonic molecular gas jet were impulsively aligned [73–75] using a stretched ($\approx 50\,fs$) linearly polarized (along z) laser pulse split off from the output of a $5\,kHz$ Titanium-Sapphire laser amplifier system, with the intensity adjusted by an aperture such that negligible $C_2H_2^+$ is observed. The remaining portion of the laser output ($\approx 90\%$) was, just like in the experiments described in Sect. 3.2, spectrally broadened in a $1\,m$ long hollow-core glass capillary, filled with neon, and the pulses were subsequently recompressed by several bounces from chirped mirrors to a duration of $\approx 4.5\,fs$ (FWHM). The duration and the CEP of each few-cycle laser pulse was measured on a single-shot basis by a stereo-ATI phasemeter [61, 62]. The short pulses, time-delayed to the aligning pulses by ΔT using a piezo stage and linearly polarized along z, ionize the C_2H_2 molecules and the three-dimensional momentum vectors of the resulting ions and ionic fragments of a single molecule were measured by cold target recoil ion momentum spectroscopy [50, 51]. The peak intensity of the ionizing laser pulses on target was determined from separate calibration measurements using single ionization of argon atoms in circularly polarized light [76].

3.3.2 Ionization Into Binding States

By the application of the alignment pulse a rotational wavepacket is created. A signature of the temporal evolution of this wavepacket derived from the ejected protons, induced by the delayed ionizing laser pulse, is shown up to the full revival at $\Delta T \approx 14\,ps$ in Fig. 3.5a (see Figure caption for details). At $\Delta T \approx 0.08\,ps$ we observe a peak due to the prompt molecular alignment, followed by the quarter- and half-revival structures of the rotational wavepacket at $\Delta T \approx 3.5\,ps$ and $\Delta T \approx 7\,ps$, respectively. The delay time where the molecules are anti-aligned with respect to the polarization direction of the alignment pulse as calculated from the molecular rotational constant is $7.07\,ps$ measured from the prompt alignment peak. Thus, anti-alignment is expected at $\Delta T = 7.15\,ps$. Indeed, we observe minimum proton signal in Fig. 3.5a at $\Delta T = 7.15\,ps$ and a maximum at $\Delta T = 6.8\,ps$, corresponding to molecules anti-aligned and aligned relative to the parallel polarization directions of the alignment and probe pulses, respectively.

Figure 3.5c shows the normalized yields of $C_2H_2^+$ and $C_2H_2^{++}$ around the half-revival structure of the rotational wavepacket at $\Delta T \approx 7\,ps$ for a laser peak intensity of the ionizing pulse of $4 \times 10^{14}\,W/cm^2$. At this intensity double ionization happens dominantly sequentially during two independent field ionization steps. In order to observe $C_2H_2^+$ ions in our experiment, after single ionization the cation must be in a binding, non-dissociative state. To reach the binding ground state of $C_2H_2^+$ the

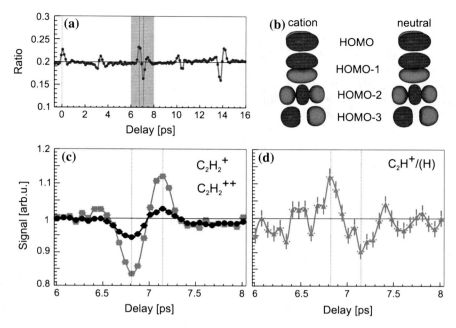

Fig. 3.5 a Ratio of the yield of protons emitted from acetylene towards the detector and the total ion yield as a function of delay between a weak, non-ionizing alignment pulse (≈50 fs) and a probe pulse (4.5 fs) measured at a probe pulse intensity of 4×10^{14} W/cm^2. The structures corresponding to prompt alignment, quarter-revival and half-revival of the rotational wavepacket created by the alignment pulse are visible at delays of 0.08, ≈3.5 and ≈7 ps, respectively. **b** The shapes of the highest four orbitals of the neutral and cation of acetylene. At the precision level of this 3D representation the orbitals of the neutral and cation look identical. **c** Yield of $C_2H_2^+$ (*dots*) and $C_2H_2^{++}$ (*squares*) as a function of delay between alignment pulse and probe pulse around the half-revival of the rotational wavepacket, normalized to the respective yields at a delay of 5.5 ps. **d** Yield of the dissociation channel $C_2H_2^+ \rightarrow C_2H^+ + H$ as a function of delay between alignment pulse and probe pulse around the half-revival of the rotational wavepacket, normalized to the yield at a delay of 5.5 ps

electron must be removed from the HOMO (highest occupied molecular orbital) or the degenerate HOMO-1 orbital of the neutral C_2H_2. Because of the shape and symmetry of these orbitals with odd parity and vanishing probability density along the molecular axis (Fig. 3.5b), molecular tunneling theory predicts a minimum of the angular ionization probability along the molecular axis and a maximum perpendicular to it [2, 3]. Accordingly, in our experiment we measure a minimum of the normalized $C_2H_2^+$ yield at $\Delta T = 6.8$ ps after the alignment pulse, where the molecules are aligned parallel to the laser polarization direction, and a maximum at $\Delta T = 7.15$ ps, where the molecules are aligned perpendicular to the laser polarization direction (Fig. 3.5c). Likewise, to observe $C_2H_2^{++}$ ions, the dication needs to be left in the ground state after the second ionization step. This is achieved by removing an electron from the HOMO of the neutral during the first ionization step and a second electron from the HOMO-1 of the cation during the second ionization step. Both involved orbitals feature maximum ionization probability perpendicular to the molecular axis. Correspondingly, we measure a much larger normalized $C_2H_2^{++}$ yield for molecules aligned perpendicular to the laser polarization direction than for molecules aligned parallel to it (Fig. 3.5c). A simulation using molecular tunneling theory [2] confirms that for the laser parameters of our experiment the first electron is removed dominantly from the HOMO (or HOMO-1) of the neutral and the second electron from the HOMO-1 of the cation. Hence, the data in Fig. 3.5c show that alignment can serve as a knob to control the preferential removal of an electron from a certain orbital during laser tunnel ionization.

3.3.3 Controlling Dissociation from the Cation

In the following we will demonstrate that the findings on $C_2H_2^+$ and $C_2H_2^{++}$ open up the opportunity of controlling the probability to fragment along a certain pathway. Moreover, it may become possible to suppress fragmentation along one pathway, while enhancing the fragmentation along another pathway. To do this we need to be able to determine in which dissociative electronic state the ion is left after ionization by controlling the ionization probability from certain orbitals. We start to investigate this opportunity using the example of the dissociation reaction $C_2H_2^+ \rightarrow C_2H^+ + H$ that starts from the cation. The measured probability of this dissociation reaction as a function of ΔT around the half-revival structure of the rotational wavepacket is shown in Fig. 3.5d. The measured probability to dissociate along this pathway is maximum for parallel ($\Delta T = 6.8\,\mathrm{ps}$) and minimum for perpendicular alignment ($\Delta T = 7.15\,\mathrm{ps}$), in opposition to the alignment dependence of creating the cation, $C_2H_2^+$, from where the dissociation reaction starts (cf. Fig. 3.5c). This is because to initiate the dissociation reaction, the first excited state of $C_2H_2^+$ needs to be populated by removing an electron from the HOMO-2 of the neutral rather than from the HOMO (or HOMO-1). Due to the structure of the HOMO-2 orbital (see Fig. 3.5b) this is preferentially achieved for parallel alignment. It is thus possible to enhance the dissociation reaction at the expense of creating $C_2H_2^+$ by changing the alignment from perpendicular to parallel, i.e. to change the probe delay from 7.15 to 6.8 ps.

3.3.4 Controlling Fragmentation Reactions from the Dication

We now turn to the investigation whether the same control strategy can be applied to enhance or suppress fragmentations from the dication that involve the creation of two ionic fragments. Figure 3.6 shows the measured fragmentation yields of the symmetric fragmentation channel, $C_2H_2^{2+} \rightarrow CH^+ + CH^+$, and the fragmentation channel involving restructuring to the vinylidene configuration prior to fragmentation by a proton migration reaction, $C_2H_2^{2+} \rightarrow CH_2^+ + C^+$, as a function of delay

Fig. 3.6 Yield of the dissociation channels $C_2H_2^{2+} \rightarrow CH_2^+ + C^+$ (**a**) and $C_2H_2^{2+} \rightarrow CH^+ + CH^+$ (**b**) as a function of delay after the application of the alignment pulse, induced around the half-revival of the rotational wavepacket by a pulse with a duration of 4.5 fs and an intensity of 4×10^{14} W/cm^2. The yield of each of the two channels has been normalized to its respective yield at a delay of 5.5 ps, when the molecules are randomly aligned

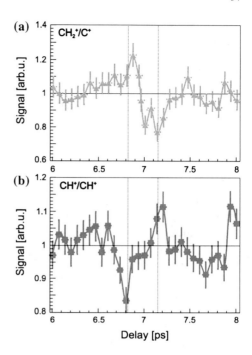

time of the ionizing probe pulse around the half-revival structure of the rotational wavepacket. First, we focus on the fragmentation reaction involving the proton migration. Figure 3.6a shows that the measured fragmentation probability for this pathway is higher for parallel ($\Delta T = 6.8$ ps) than for perpendicular ($\Delta T = 7.15$ ps) alignment. To explain the observed alignment dependence we propose the following scenario: During the first ionization step an electron is removed from the HOMO of the neutral, and during the second ionization step an electron from the HOMO-2 (or HOMO-3) is removed. Preliminary simulations based on 3D time-dependent density functional theory (TDDFT) using the OCTOPUS code [77–79] show that the angular dependence of the ionization probability into the dication resulting from two (independent) sequential ionization steps at the laser parameters used in the experiments peaks along the laser polarization direction, in agreement with the experiment. The TDDFT simulations were performed within a frozen nuclei approximation. Absorbing boundaries were used to collect the electronic density leaving the molecule. The norm reduction of each Kohn-Sham orbital after the end of the pulse is interpreted as the ionization probability from this orbital. If strong mixing between orbitals occurred due to the action of the laser pulse, those orbitals were treated together. Comparison with the experiment was achieved by convolving the numerically obtained angle dependent ionization probability with the alignment distribution of a simulated rotational wavepacket [74, 75, 80] induced by a pulse with the same parameters as in the experiment, and for an initial rotational temperature of 100 K that matches the experimental conditions.

A more intuitive explanation for the experimentally observed fragmentation dependence can be obtained by considering that the fragmentation reaction takes predominately place on the excited cationic electronic surface $^3\Pi_u$ [81]. This state is populated by removing an electron from the HOMO of the neutral during the first ionization step, and an electron from the HOMO-2 during the second ionization step. Removal of an electron from the HOMO of the neutral is preferentially achieved for perpendicular alignment. However, for the laser intensity used here the dominant ionization step during which it is decided into which state of $C_2H_2^{++}$ ionization takes place is the second one. Thus, to make the molecule fragment preferentially along the pathway $C_2H_2^{2+} \rightarrow CH_2^+ + C^+$ by removing an electron from the HOMO-2, the molecule needs to be aligned parallel to the laser polarization direction. For perpendicular alignment, ionization will dominantly occur from the HOMO and HOMO-1 and fragmentation will preferentially take place via other pathways, leading to minimal fragmentation into CH_2^+/C^+. The experimental data in Fig. 3.6a are in perfect agreement with these explanations.

Now we turn to the fragmentation pathway $C_2H_2^{2+} \rightarrow CH^+ + CH^+$, where the molecule fragments via breakage of the center bond. In the experiment, fragmentation along this pathway is maximized for perpendicular alignment and minimized for parallel alignment (Fig. 3.6b). To enhance (or suppress) fragmentation along this pathway the population of the surface $^3\Pi_u$ of $C_2H_2^{++}$ needs to be controlled [81]. As discussed above, this is achieved by removing an electron from the HOMO of the neutral during the first ionization step, and an electron from the HOMO-2 during the second ionization step. For fragmentation by breakage of the center bond on the $^3\Pi_u$ surface a relatively deep minimum of roughly $1.5\,eV$ needs to be overcome [81]. Therefore, we propose that the $^3\Pi_u$ surface in our experiment is populated dominantly by recollision double ionization. This is motivated by the following two facts: Firstly, recollision double ionization leads to a population of the $^3\Pi_u$ surface at a considerably higher energy than by ionization within the Franck-Condon region, as for double ionization by recollision the time difference between the first and second ionization step is less than $2\,fs$. Within this short time the CH^+ fragments do not move much apart leading to a population of the surface at a higher potential energy. Secondly, recollision ionization starts with ionization from the HOMO of the neutral. Therefore, neglecting the angular dependence of the recollision cross section, the alignment dependence will be dominated by the structure of the HOMO of the neutral, in agreement with the experimental data in Fig. 3.6b.

3.3.5 Selecting the Fragmentation Pathway

In a molecular fragmentation a final set of reaction products can, in general, not only be obtained by fragmentation from a single, but also from several initial states, see e.g. [82]. As the here demonstrated method for controlling a fragmentation reaction directly addresses the initial electronic state, it becomes possible to not only determine the final reaction products, but also the path along which the molecule

3 Exploring and Controlling Fragmentation

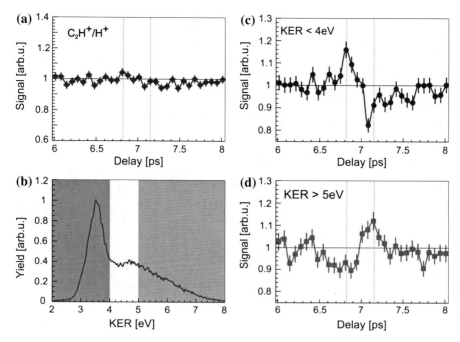

Fig. 3.7 **a** Yield of the dissociation channel $C_2H_2^{2+} \to C_2H^+ + H^+$ as a function of delay after the application of the alignment pulse, probed around the half-revival of the induced rotational wavepacket by a pulse with a duration of 4.5 fs and an intensity of 4×10^{14} W/cm^2, normalized to the yield at a delay of 5.5 ps, when the molecules are randomly aligned. **b** Spectrum of the kinetic energy release (KER) of the fragments C_2H^+ and H^+. **c**, **d** Yield of the dissociation channel shown in (**a**) when the KER is restricted to < 4 eV (**c**) and > 5 eV (**d**)

fragments. We will demonstrate that in the following for the well studied deprotonation pathway of acetylene, $C_2H_2^{++} \to C_2H^+ + H^+$. The measured yield of this pathway as a function of probe delay shows no apparent dependence on the transient alignment of the molecular axis (Fig. 3.7a). However, the kinetic energy release (KER) spectrum exhibits a bimodal shape (Fig. 3.7b), indicating that the deprotonation dynamics may proceed along two distinct fragmentation pathways. Using the coincidence capabilities of our experiment we can decompose the fragmentation yield of this pathway into the contributions of the narrow low energy (KER < 4 eV) and the broad high energy (KER > 5 eV) peaks. The resulting fragmentation yields from the two KER-regions feature a strong opposite dependence on the molecular alignment (Fig. 3.7c, d): While the fragmentation probability of molecules with high KER is higher for perpendicular alignment, lower energy fragmentation is maximized for parallel alignment. We demonstrate in Fig. 3.8 that the fragmentations leading to higher KER are triggered by recollision double ionization, while the fragmentations associated with the lower KER are most likely triggered by field ionization. This can be seen clearly from both the pulse duration (Fig. 3.8a) and intensity (Fig. 3.8b) dependence of the KER spectra. Especially the very strong decrease of the high KER

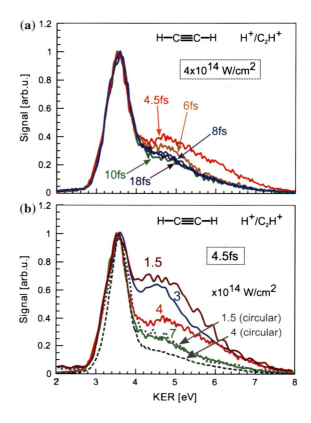

Fig. 3.8 Pulse duration **a** and intensity dependence **b** of the kinetic energy release (KER) spectrum of the fragmentation channel $C_2H_2^{2+} \rightarrow C_2H^+ + H^+$. The spectra in **a** are all for the same intensity of 4×10^{14} W/cm^2. The spectra in **b** are all for the shortest pulse duration of 4.5 fs. The intensities and pulse durations are indicated in the figures

spectral parts with increasing intensity is a strong indication that this spectral part is caused by recollision double ionization. In addition, the spectra obtained with circular polarization, for which electron recollisions are suppressed, show a very strong decrease of the high energy KER. In contrast, the low energy part of the KER spectra does neither change with pulse duration nor with intensity. Having identified that the low energy fragments originate from sequential field double ionization, while the fragmentations leading to high KER are triggered by recollision double ionization, it is now easy to explain the opposite dependence of the yield on the transient alignment for the two paths (Fig. 3.7): Sequential field ionization involves removal of an electron from the HOMO-2 or HOMO-3 of the cation which show negligible probability density perpendicular to the molecular axis. Accordingly, the fragmentation yield for low KER shows a maximum for parallel alignment and a minimum for perpendicular alignment. The probability of recollision double ionization, as discussed above, peaks for perpendicular alignment, because it starts from the HOMO of the neutral showing maximum probability density perpendicular to the molecular axis.

3.3.6 Discussion and Outlook

We have shown, using the acetylene molecule as an example, that molecular fragmentations can be controlled by the selective removal of electrons from certain (lower-valence) orbitals. Our experiments demonstrate that this can be achieved with considerable selectivity using the alignment of the molecular axis with respect to the laser polarization direction. In Sect. 3.2 we have discussed that the fragmentation of polyatomic molecules can be controlled by the selective population of excited dissociative states due to recollision-ionization from lower-valence orbitals. The control mechanism described in this section is based on exactly the same idea, only the control parameter is different (alignment vs. CEP). An important feature is that both control schemes are based on purely electronic transitions and can thus be applied on attosecond time-scales, e.g. by electron recollision. In the future, both control methods, CEP and alignment, may be applied together, thereby possibly leading to an enhanced selectivity allowing to selectively control the fragmentation of one pathway only.

3.4 Exploring Many-Electron Ionization Dynamics in Polyatomic Molecules

In this section, we report on experiments aimed at investigating the coupled electronic and nuclear dynamics behind laser-ionization leading to multiply ionized hydrocarbon molecules [39, 83–85], and the possibility of extending our understanding of enhanced ionization (EI) [24, 25] to polyatomic molecules. Specifically we study ionization of acetylene, C_2H_2, and ethylene, C_2H_4, molecules subject to near-infrared ($\sim$750 nm) laser pulses. As EI involves nuclear motion, it can be suppressed by using pulse durations shorter than the time needed for the involved bonds to stretch to the critical internuclear distance [32, 34, 86], R_c, at which EI takes place. In our investigations we therefore employ pulses with durations, τ (FWHM), ranging from the few-cycle ($\sim$4.5 fs) to the multi-cycle (27 fs) regime exhibiting laser peak intensities in the range 2 to 8×10^{14} W/cm^2.

We are particularly interested in the ionization dynamics behind the surprisingly high charge states observed in [39] for a series of hydrocarbon molecules (CH_4, C_2H_4, C_4H_6, C_6H_{14}). In that paper it was shown that with 27 fs laser pulses at 790 nm very high charge states (e.g. $\geq +13$ for C_4H_6) could be reached already at moderate intensities of a few 10^{14} W/cm^2. These high charge states result in complete fragmentation of the molecular ions and the emission of protons with considerable kinetic energy. Based on experimental evidence it was suggested that a novel type of molecular ionization, i.e. a multi-bond version of the EI mechanism taking place in parallel at many C–H bonds, be responsible for the surprising ionization behaviour.

Simulations based on time-dependent density functional theory (TD-DFT) performed for CH_4 and C_4H_6 have confirmed the experimental findings that, as a result of the interaction with the laser pulse, the molecules completely disintegrate into atomic ions via a concerted all-at-once Coulomb explosion resulting in remarkably high and similar proton energies [83]. However, while the simulations could confirm the fragmentation dynamics and the origin of the observed high proton energies observed in the experiments [39], the proposed multi-bond EI mechanism could not be conclusively confirmed.

A recent theoretical study [84, 85] used time-dependent Hartree-Fock (TDHF), restricted to one spatial dimension, to investigate the ionization mechanism that leads to the experimentally observed high charge states. The results obtained for a model potential of the acetylene molecule, C_2H_2, with fixed nuclei clearly reveal the existence of a critical C–H internuclear distance, R_c, at which the ionization rate is strongly enhanced. When the C–H distance is adjusted to the critical distance, the model predicts that at a laser intensity of 14×10^{14} W/cm^2 ionization proceeds not only from the HOMO but also from HOMO-1 and HOMO-2, and that the charge densities from the HOMO and HOMO-1 efficiently localize at the positions of the protons—as it is required for the EI mechanism. In addition, the laser field efficiently couples the HOMO-2 to the two higher lying orbitals such that in total about 6 electrons are removed from the C_2H_2 molecule. Thus, the TDHF simulations support the multi-bond EI mechanism proposed by Roither et al. [39].

3.4.1 Experiment

The experiments are similar to those described in Sects. 3.2 and 3.3. In short, few-cycle laser pulses are generated by spectral broadening of 25 fs laser pulses with a spectrum centered at 790 nm from a Titanium-Sapphire laser amplifier system in a 1 m long hollow-core glass capillary filled with neon at several atmospheres overpressure. Temporal recompression of the pulses after the capillary by several bounces from pairs of chirped mirrors results in a pulse duration of $\sim$4.5 fs. The duration and intensity-stability of the pulses are monitored on a shot-to-shot basis by a stereo-ATI phase-meter [61, 62]. The pulses are directed into the cold target recoil ion momentum spectroscopy (COLTRIMS) [50] apparatus described above, where they are focused by a spherical mirror with a focusing length of 60 mm into a supersonic gas jet of randomly aligned ethylene or acetylene molecules. The resulting ionic fragments from the interaction of a single molecule with a single laser pulse are guided over a length of 5.7 cm to a position and time sensitive detector by a weak homogeneous electric field (9.4 V/cm) where the positions and times of flight of the fragments are recorded in coincidence. From this information the three-dimensional momentum of each ionic fragment is calculated.

3 Exploring and Controlling Fragmentation

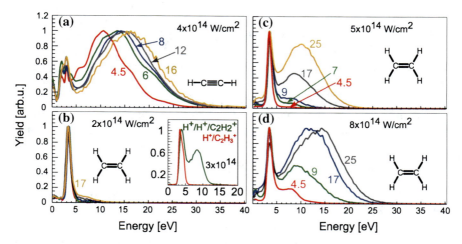

Fig. 3.9 Energy spectra of protons ejected from acetylene (**a**) and ethylene (**c**–**d**) during complete fragmentation with pulses whose intensities and durations are indicated in the figures: For each of the four spectral series the same intensity has been used, as indicated in each panel. Each spectrum is labeled with a pulse duration (FWHM) in fs. The inset in **b** shows for comparison the proton energy spectra of two fragmentation channels (as indicated), where the fragmentations start from doubly and triply charged ethylene. The intensity of the used pulse is indicated, its duration was 4.5 fs

3.4.2 Proton Spectra: Dependence on Pulse Duration and Intensity

From the recorded coincidence momenta of fragment ions we extracted background-free kinetic energy spectra of the protons that are created during the laser-molecule interaction by selecting all protons that are emitted towards the detector within a 90° cone [39]. The proton energy spectra are shown in Fig. 3.9 for both C_2H_2 and C_2H_4 as a function of pulse duration and different pulse peak intensities. The peak intensity was adjusted by reflecting the beam off a fused silica (FS) block under different angles, thereby changing the pulse energy content. Calibration of the peak intensity on target was done with an estimated precision of ±10 % by separate measurements using single ionization of argon atoms in circularly polarized light [76]. The pulse duration was varied by positively chirping the shortest pulses by propagating them through different amounts of fused silica. We checked by remeasuring some of the spectra with negatively chirped pulses, that the results presented below are independent of the sign of the chirp of the pulses. Furthermore, as can be seen in Fig. 3.9, in the multi-cycle pulse limit ($\tau \gtrsim 17$ fs) the results obtained with Fourier limited pulses of $\tau = 27$ fs are almost recovered.

Figure 3.9 shows that for a given laser intensity for both C_2H_2 and C_2H_4 the proton energy spectra extend to higher energies with increasing pulse durations. While for the shortest pulse duration of 4.5 fs, for C_2H_4 the proton energies do not significantly exceed 10 eV even for the higher intensity of 8×10^{14} W/cm², the proton energies extend up to 25 eV when the pulse duration is increased to 17 fs. Further increasing the pulse duration to 27 fs does not lead to significantly higher proton energies. Using coincidence selection we extract the proton energy spectra created by Coulomb explosion of doubly and triply charged ethylene ions during two- and three-body fragmentations (inset of Fig. 3.9b). By comparing these spectra to those measured for the 4.5 fs pulse in Fig. 3.9d it becomes evident that even at 8×10^{14} W/cm^2 the low proton energies are created dominantly by two- and three-body fragmentations from low charge states. Only the higher proton energies observed for longer pulses are exclusively created by complete decompositions of the C_2H_4 molecules into 6 ionic fragments from high charge states.

A detailed coincidence analysis of the data corresponding to the spectra in Fig. 3.9c, d shows that protons with an energy of $\gtrsim 12$ eV are created during complete fragmentation of the ethylene molecule from charge states higher or equal of +6. Protons with less energy are ejected during incomplete fragmentations from lower charge states. From the spectra in Fig. 3.9c, d it can, thus, be inferred that only for pulse durations $\tau \gtrsim 10$ fs a significant amount of molecules is ionized to charge states higher than +6. For pulse durations shorter than 10 fs complete fragmentation of the ethylene molecules is negligible even for the highest intensity. For long pulses around 20 fs complete molecular fragmentations are observed even for intensities as low as 2×10^{14} W/cm^2, and for 8×10^{14} W/cm^2 the probability to completely fragment the molecule from a high charge states is higher than 50%. These data seem to suggest the existence of a mechanism that leads to a strong increase in multi-electron ionization probability for $\tau \gtrsim 10$ fs, and little probability for ionization to high charge states for shorter pulses. A possible conclusion, hence, would be that these data clearly confirm the multi-bond EI mechanism proposed in [39], as, seemingly, only for $\tau \gtrsim 10$ fs the stretch motion of the C–H bonds to the critical internuclear distance can be completed before the trailing edge of the laser pulse. For shorter pulses the R_c of the C–H bonds cannot be reached before the pulse fades and the ionization proceeds less efficiently. While this interpretation would be in agreement with experimental findings for the H_2 molecule [32], we will show below that this conclusion is, however, premature and inconsistent with a more involved coincidence data analysis presented in Fig. 3.11.

We now turn to the proton spectra observed for acetylene, C_2H_2 (Fig. 3.9a). They show a similar dependence on pulse duration as those for ethylene. However, the proton energies for C_2H_2 at a given pulse duration are consistently higher than those for ethylene, even though the intensity used for C_2H_2 is smaller. Lötstedt et al. have pointed out [84, 85] that the data presented in [39] are consistent with an interpretation that, regardless of the molecular species, the highest proton energies are created by Coulomb explosions from charge states where on average $2n$ electrons are ejected from the molecules, with n the number of C–H bonds in the molecule. Considering that for C_2H_4 $2n = 8$ and for C_2H_2 $2n = 4$, and taking into account that the energy released during the Coulomb explosion from a given charge state is shared amongst the participating n protons, the proton energies for a given charge state are thus expected to be similar for C_2H_4 and C_2H_2. The proton spectra in Fig. 3.9 thus suggest that the ionization in C_2H_2 proceeds more efficiently than in C_2H_4.

3 Exploring and Controlling Fragmentation

Fig. 3.10 Energy spectra of protons ejected during complete fragmentation of acetylene ions with a certain charge state (Z, as indicated in the figures) initiated with pulses whose intensities and durations are indicated in the figures. The spectra are the results of a four-body coincidence selection of two protons and two carbon ions C^{n+} and C^{m+}, when the charge state $Z = m + n + 2$

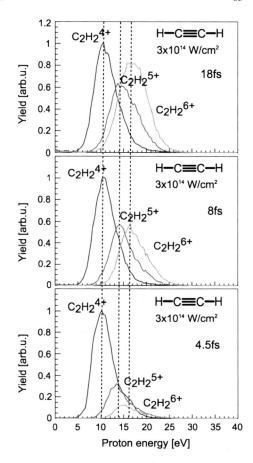

A possible explanation is that the underlying ionization mechanism might be sensitive to the alignment of the the C–H bonds with respect to the laser field direction. As for C_2H_4, as opposed to C_2H_2, not all C–H bonds can be aligned with the laser field, the ionization might proceed less efficiently for C_2H_4 than for C_2H_2.

3.4.3 Charge State Selected Proton Spectra and Reconstruction of the C–H Distance

In order to obtain insight into the ionization dynamics that lead to the multiply charged molecules and subsequently to the energetic protons, we performed a complete four-body coincidence analysis of the fragments from acetylene. The proton spectra corresponding to fragmentation of C_2H_2 from charge states $Z = +4$ to $Z = +6$ resulting from this four-body coincidence selection are plotted in Fig. 3.10

Fig. 3.11 a Mean energy of protons ejected during complete fragmentation of acetylene ions with a certain charge state (Z, as indicated for the three uppermost data series) as a function of pulse duration and a constant intensity of 3×10^{14} W/cm^2. The two lower data series, shown for comparison, are for protons ejected during breakage into three fragments. **b** C–H distance at the time of proton ejection calculated from the proton energies in (**a**), shown with the same data point style (see text for details)

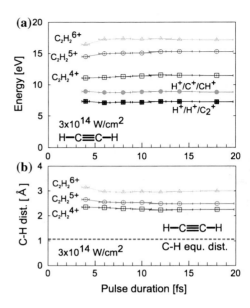

for different pulse durations and intensities. We notice in the energy spectra of the protons measured for 3×10^{14} W/cm^2, that the two protons ejected from a certain charge state Z show a smooth shape and are centered around a certain energy that increases with Z. These features are consistent with a concerted Coulomb explosion dynamics of both protons sensing a very similar positive charge Z [39]. However, the two most important features of these spectra are that (i) the proton energies corresponding to a certain charge state are roughly independent of the pulse duration, and (ii) the *yield* of fragmenting from a certain charge state monotonically increases with pulse duration.

Figure 3.11a shows a comprehensive overview over the proton energies, calculated as the mean value of proton spectra of the type shown in Fig. 3.10, for the charge states $Z = +4$ to $Z = +6$ and for two fragmentation channels starting from $Z = +3$ as a function of pulse duration. It can be seen that the proton energies are almost independent of the pulse duration for a given charge state, even down to the shortest pulse duration of 4.5 fs. Under the assumption that the fragmentation reactions following ionization are driven by pure Coulomb repulsion, from the proton energies the C–H distance at the time when the Coulomb explosion starts can be calculated. The results of this calculation are shown in Fig. 3.11b: The C–H distance is independent of the laser pulse duration and always larger than the equilibrium C–H distance. In order to calculate the C–H distance one needs to assume a certain charge density distribution along the C–C structure, as the charge distribution within the molecule is not known. For the calculation of the values shown in Fig. 3.11b we assumed for simplicity that the charge on the C–C structure, $Z - 2$, is concentrated in the center. Then, the proton energy, from which the C–H distance $R_{\text{C–H}}$ can be calculated, reads

$$E = \frac{Z-2}{R_{C-H} + 1/2R_{C-C}} + \frac{1}{4(R_{C-H} + 1/2R_{C-C})},$$

with $R_{C-C} = 1.2 \, \overset{\circ}{A}$ the equilibrium distance of the C–C bond. Although, the such calculated C–H distance is only an estimate, this does not affect the conclusion that the C–H distance where the Coulomb explosion starts is independent of the pulse duration for all charge states. The two facts revealed by Fig. 3.11b, namely that (i) the estimated C–H distance is roughly 2.5 times larger than the equilibrium C–H distance, and (ii) that the C–H distance is the same for all pulse durations even for the shortest ones, can be interpreted as follows: Firstly, these facts allow for a conclusion that the ionization process indeed happens via enhanced ionization as suggested previously [39, 84, 85]. During the onset of the laser pulse the molecules become ionized which initiates the C–H stretch motion to the critical internuclear distance R_c where ionization takes place multiply times within a short period. Secondly, the independence of the C–H distance on the pulse duration means that the C–H stretch motion to R_c takes place very quickly and cannot be suppressed even with pulse durations as short as 4.5 fs.

3.4.4 Dependence of Charge State on Pulse Duration

The results of our experiments, thus, suggest that there indeed exists a multi-bond version of the EI mechanism in polyatomic molecules, as first proposed by Roither et al. [39] and later theoretically confirmed by Lötstedt et al. [84, 85] using simulations with fixed C–H distances. As EI seems to take place for all pulse durations, even for the shortest ones, how can we then explain the strong dependence of the proton energies on the pulse duration shown in Fig. 3.9? To answer this question we plot in Fig. 3.12 the yield of a certain charge state as a function of pulse duration measured for ethylene. It can be seen that the relative yield of a certain charge state increases over several orders of magnitude when the pulse duration is increased from 4.5 to 27 fs, and higher charge states show a steeper dependence on the pulse duration. The proton spectra shown in Fig. 3.9 are the results of fragmentations from a mix of charge states. For short pulse durations the contributions from high charge states is negligible and the protons are ejected dominantly from small charge states. For longer pulse durations also higher charge states contribute significantly and the protons are ejected, on average, from a higher charge state leading to higher proton energies.

3.4.5 Discussion and Outlook

The results of the experiments described here, aimed at investigating the mechanism behind the surprisingly high charge states observed in the ionization of a series

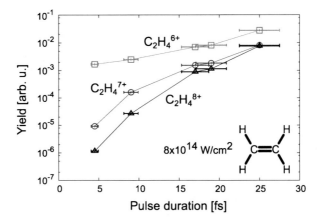

Fig. 3.12 Yield of three complete fragmentation channels starting from charge states $Z = 6, 7, 8$ as a function of pulse duration measured at the indicated intensity

of hydrocarbon molecules [39], are in agreement with a multi-bond version of EI, although they do not offer conclusive evidence based on a pulse duration dependence of the C–H distance from which the fragmentations occur. In our experiments the C–H distance is almost independent of the pulse duration, which means that the time for the C–H bonds to stretch to the critical internuclear distance is sufficient even for pulse durations as short as 4.5 fs. Nevertheless, we observe a strong dependence of the yield of a certain charge state on the pulse duration. For short pulse durations the molecules are mainly put to lower charge states, only for longer pulse durations the molecules are efficiently put to high charge states. This pulse duration dependence of the yield allows explaining the observed proton spectra in Fig. 3.9. Our experiments, however, do not give insight into the origin of the strong dependence of the yield. Future work should thus be dedicated to investigating the coupled electronic and nuclear dynamics underlying this yield dependence. A possible experimental route to its investigation could be the exploitation of molecular alignment. Another route would be probing the ionization and fragmentation behaviour with still shorter, sub-femtosecond, pulses.

3.5 Summary

In this chapter, based on the results of experiments, we discussed the role of the sub-cycle-triggered removal of valence electrons by strong-field ionization with few-cycle laser pulses for the fragmentation of polyatomic molecules. Starting in Sect. 3.2 from the observation that the fragmentation yield of various fragmentation channels in acetylene, C_2H_2, ethylene, C_2H_4, and 1,3-butadiene, C_4H_6, strongly depends on the sub-cycle shape of few-cycle laser pulses, as defined by the carrier-envelope phase (CEP), we found that this dependence is caused by a strongly

CEP-dependent probability of removing an electron from inner-valence shells by recollision ionization. As electron removal from inner-valence shells puts the molecular ion in many cases into an excited dissociative state, this result opens up the possibility to control molecular fragmentation reactions by intense laser fields. We argued that the controlled population of an excited dissociative state by lower-valence ionization is a general method for controlling, i.e. enabling or suppressing, fragmentation reactions. Indeed, we could show in Sect. 3.3 that fragmentation control can also be achieved by using the alignment of the molecular axis with respect to the laser polarization direction as a control parameter. In this case the different angular tunnel ionization probability for different (lower-valence) orbitals is exploited for gaining selectivity in removing electrons from specific orbitals. Electron removal by strong-field ionization is a process that takes place on sub-cycle times-scales. Thus, a controlled ionization to dissociative excited states can be used for controlling molecular fragmentation reactions on the attosecond time-scale. On this time-scale nuclear motion is usually irrelevant. However, nuclear motion taking place during an (or a series of) ionization event(s) may lead to a substantial modification of the ionization dynamics and therefore, in the general case, needs to be taken into account. In Sect. 3.4 we presented experiments that address this coupling between the nuclear and electronic dynamics by a systematic investigation of the dependence of the fragmentation products on the pulse duration and intensity of the ionizing laser pulses. We found that for the here investigated hydrocarbon molecules the nuclear motion, in particular the C–H stretch motion, leads to a significant modification of the ionization behaviour as compared to frozen nuclei, even then when the time during which ionization is possible is limited to a few femtoseconds. Our experiments, thus, further strengthen the hypothesis of the existence of a multi-bond enhanced-ionization mechanism that takes place in parallel at many C–H bonds stretched to a critical internuclear distance.

Acknowledgments Funding by the Austrian Science Fund (FWF) under grants P21463-N22, SFB049-Next Lite and I274-N16, and by the ERC within the SIRG scheme is gratefully acknowledged. We thank K. Doblhoff-Dier, S. Gräfe, H.-L. Xu, T. Rathje, G. G. Paulus, S. Bubin, K. Varga, E. Lötstedt, and K. Yamanouchi for contributions and discussions.

References

1. C. Guo, M. Li, J. Nibarger, G. Gibson, Phys. Rev. A **58**(6), R4271 (1998)
2. X. Tong, Z. Zhao, C. Lin, Phys. Rev. A **66**(3), 033402 (2002)
3. J. Muth-Böhm, A. Becker, F. Faisal, Phys. Rev. Lett. **85**(11), 2280 (2000)
4. D. Pavičić, K. Lee, D. Rayner, P. Corkum, D. Villeneuve, Phys. Rev. Lett. **98**(24), 243001 (2007)
5. A. Alnaser, S. Voss, X. Tong, C. Maharjan, P. Ranitovic, B. Ulrich, T. Osipov, B. Shan, Z. Chang, C. Cocke, Phys. Rev. Lett. **93**(11), 113003 (2004)
6. D. Mathur, A. Dharmadhikari, F. Rajgara, J. Dharmadhikari, Phys. Rev. A **78**(1), 013405 (2008)
7. L. Holmegaard, J.L. Hansen, L. Kalhøj, S. Louise Kragh, H. Stapelfeldt, F. Filsinger, J. Küpper, G. Meijer, D. Dimitrovski, M. Abu-samha, C.P.J. Martiny, L. Bojer Madsen, Nat. Phys. **6**(6), 428 (2010)

8. J.L. Hansen, L. Holmegaard, J.H. Nielsen, H. Stapelfeldt, D. Dimitrovski, L.B. Madsen, J. Phys. B: At. Mol. Opt. Phys. **45**(1), 015101 (2012)
9. A. Talebpour, A. Bandrauk, J. Yang, S. Chin, Chem. Phys. Lett. **313**(5–6), 789 (1999)
10. B.K. McFarland, J.P. Farrell, P.H. Bucksbaum, M. Gühr, Science (New York, N.Y.) **322**(5905), 1232 (2008)
11. P. Hoff, I. Znakovskaya, S. Zherebtsov, M.F. Kling, R. Vivie-Riedle, Appl. Phys. B **98**(4), 659 (2009)
12. H. Akagi, T. Otobe, A. Staudte, A. Shiner, F. Turner, R. Dörner, D.M. Villeneuve, P.B. Corkum, Science (New York, N.Y.) **325**(5946), 1364 (2009)
13. C. Wu, H. Zhang, H. Yang, Q. Gong, D. Song, H. Su, Phys. Rev. A **83**(3), 033410 (2011)
14. A.E. Boguslavskiy, J. Mikosch, A. Gijsbertsen, M. Spanner, S. Patchkovskii, N. Gador, M.J.J. Vrakking, A. Stolow, Science (New York, N.Y.) **335**(6074), 1336 (2012)
15. M. Lezius, V. Blanchet, D. Rayner, D. Villeneuve, A. Stolow, M. Ivanov, Phys. Rev. Lett. **86**(1), 51 (2001)
16. M. Lezius, V. Blanchet, M.Y. Ivanov, A. Stolow, J. Chem. Phys. **117**(4), 1575 (2002)
17. A. Markevitch, S. Smith, D. Romanov, H. Bernhard Schlegel, M. Ivanov, R. Levis, Phys. Rev. A **68**(1), 011402(R) (2003)
18. I. Litvinyuk, F. Légaré, P. Dooley, D. Villeneuve, P. Corkum, J. Zanghellini, A. Pegarkov, C. Fabian, T. Brabec, Phys. Rev. Lett. **94**(3), 033003 (2005)
19. H.W. van der Hart, J. Phys. B: At. Mol. Opt. Phys. **33**(20), L699 (2000)
20. V.L.B.D. Jesus, B. Feuerstein, K. Zrost, D. Fischer, A. Rudenko, F. Afaneh, C.D. Schröter, R. Moshammer, J. Ullrich, J. Phys. B: At. Mol. Opt. Phys. **37**(8), L161 (2004)
21. A. Rudenko, K. Zrost, B. Feuerstein, V. de Jesus, C. Schröter, R. Moshammer, J. Ullrich, Phys. Rev. Lett. **93**(25), 253001 (2004)
22. T. Ikuta, K. Hosaka, H. Akagi, A. Yokoyama, K. Yamanouchi, F. Kannari, R. Itakura, J. Phys. B: At. Mol. Opt. Phys. **44**(19), 191002 (2011)
23. X. Xie, K. Doblhoff-Dier, S. Roither, M.S. Schöffler, D. Kartashov, H. Xu, T. Rathje, G.G. Paulus, A. Baltuška, S. Gräfe, M. Kitzler, Phys. Rev. Lett. **109**(24), 243001 (2012)
24. T. Seideman, M. Ivanov, P. Corkum, Phys. Rev. Lett. **75**(15), 2819 (1995)
25. T. Zuo, A.D. Bandrauk, Phys. Rev. A **52**(4), R2511 (1995)
26. L. Frasinski, K. Codling, P. Hatherly, J. Barr, I. Ross, W. Toner, Phys. Rev. Lett. **58**(23), 2424 (1987)
27. K. Codling, L.J. Frasinski, P.A. Hatherly, J. Phys. B: At. Mol. Opt. Phys. **22**(12), L321 (1989)
28. D. Strickland, Y. Beaudoin, P. Dietrich, P. Corkum, Phys. Rev. Lett. **68**(18), 2755 (1992)
29. C. Cornaggia, D. Normand, J. Morellec, J. Phys. B: At. Mol. Opt. Phys. **25**(17), L415 (1992)
30. M. Schmidt, D. Normand, C. Cornaggia, Phys. Rev. A **50**(6), 5037 (1994)
31. E. Constant, H. Stapelfeldt, P.B. Corkum, Phys. Rev. Lett. **76**(22), 4140 (1996)
32. A. Alnaser, X. Tong, T. Osipov, S. Voss, C. Maharjan, P. Ranitovic, B. Ulrich, B. Shan, Z. Chang, C. Lin, C. Cocke, Phys. Rev. Lett. **93**(18), 183202 (2004)
33. T. Ergler, B. Feuerstein, A. Rudenko, K. Zrost, C. Schröter, R. Moshammer, J. Ullrich, Phys. Rev. Lett. **97**(10), 103004 (2006)
34. I. Bocharova, R. Karimi, E. Penka, J.P. Brichta, P. Lassonde, X. Fu, J.C. Kieffer, A. Bandrauk, I. Litvinyuk, J. Sanderson, F. Légaré, Phys. Rev. Lett. **107**(6), 063201 (2011)
35. T. Weber, H. Giessen, M. Weckenbrock, G. Urbasch, A. Staudte, L. Spielberger, O. Jagutzki, V. Mergel, M. Vollmer, R. Dörner, Nature **405**(6787), 658 (2000)
36. A. Staudte, C. Ruiz, M. Schöffler, S. Schössler, D. Zeidler, T. Weber, M. Meckel, D. Villeneuve, P. Corkum, A. Becker, R. Dörner, Phys. Rev. Lett. **99**(26), 263002 (2007)
37. A. Rudenko, V. de Jesus, T. Ergler, K. Zrost, B. Feuerstein, C. Schröter, R. Moshammer, J. Ullrich, Phys. Rev. Lett. **99**(26), 263003 (2007)
38. B. Bergues, M. Kübel, N.G. Johnson, B. Fischer, N. Camus, K.J. Betsch, O. Herrwerth, A. Senftleben, A.M. Sayler, T. Rathje, T. Pfeifer, I. Ben-Itzhak, R.R. Jones, G.G. Paulus, F. Krausz, R. Moshammer, J. Ullrich, M.F. Kling, Nat. Commun. **3**(may), 813 (2012)
39. S. Roither, X. Xie, D. Kartashov, L. Zhang, M. Schöffler, H. Xu, A. Iwasaki, T. Okino, K. Yamanouchi, A. Baltuska, M. Kitzler, Phys. Rev. Lett. **106**(16), 163001 (2011)

3 Exploring and Controlling Fragmentation 71

40. M.F. Kling, C. Siedschlag, A.J. Verhoef, J.I. Khan, M. Schultze, T. Uphues, Y. Ni, M. Uiberacker, M. Drescher, F. Krausz, M.J.J. Vrakking, Science (New York, N.Y.) **312**(5771), 246 (2006)
41. M. Kling, C. Siedschlag, I. Znakovskaya, A. Verhoef, S. Zherebtsov, F. Krausz, M. Lezius, M. Vrakking, Mol. Phys. **106**(2–4), 455 (2008)
42. I. Znakovskaya, P. von den Hoff, G. Marcus, S. Zherebtsov, B. Bergues, X. Gu, Y. Deng, M. Vrakking, R. Kienberger, F. Krausz, R. de Vivie-Riedle, M. Kling, Phys. Rev. Lett. **108**(6), 063002 (2012)
43. M. Kremer, B. Fischer, B. Feuerstein, V.L.B. de Jesus, V. Sharma, C. Hofrichter, A. Rudenko, U. Thumm, C.D. Schröter, R. Moshammer, J. Ullrich, Phys. Rev. Lett. **103**(21), 213003 (2009)
44. B. Fischer, M. Kremer, T. Pfeifer, B. Feuerstein, V. Sharma, U. Thumm, C. Schröter, R. Moshammer, J. Ullrich, Phys. Rev. Lett. **105**(22), 223001 (2010)
45. I. Znakovskaya, P. von den Hoff, N. Schirmel, G. Urbasch, S. Zherebtsov, B. Bergues, R. de Vivie-Riedle, K.M. Weitzel, M.F. Kling, Phys. Chem. Chem. Phys.: PCCP **13**(19), 8653 (2011)
46. I. Znakovskaya, P. von den Hoff, S. Zherebtsov, A. Wirth, O. Herrwerth, M. Vrakking, R. de Vivie-Riedle, M. Kling, Phys. Rev. Lett. **103**(10), 103002 (2009)
47. Y. Liu, X. Liu, Y. Deng, C. Wu, H. Jiang, Q. Gong, Phys. Rev. Lett. **106**(7), 073004 (2011)
48. P. Corkum, Phys. Rev. Lett. **71**(13), 1994 (1993)
49. F. Lindner, G. Paulus, H. Walther, A. Baltuška, E. Goulielmakis, M. Lezius, F. Krausz, Phys. Rev. Lett. **92**(11), 113001 (2004)
50. R. Dörner, V. Mergel, O. Jagutzki, L. Spielberger, J. Ullrich, R. Moshammer, H. Schmidt-Böcking, Phys. Rep. **330**(2–3), 95 (2000)
51. L. Zhang, S. Roither, X. Xie, D. Kartashov, M. Schöffler, H. Xu, A. Iwasaki, S. Gräfe, T. Okino, K. Yamanouchi, A. Baltuska, M. Kitzler, J. Phys. B: At. Mol. Opt. Phys. **45**(8), 085603 (2012)
52. C. Maharjan, A. Alnaser, X. Tong, B. Ulrich, P. Ranitovic, S. Ghimire, Z. Chang, I. Litvinyuk, C. Cocke, Phys. Rev. A **72**(4), 041403 (2005)
53. C. Smeenk, J.Z. Salvail, L. Arissian, P.B. Corkum, C.T. Hebeisen, A. Staudte, Opt. Express **19**(10), 9336 (2011)
54. T.H. Dunning, J. Chem. Phys. **90**(2), 1007 (1989)
55. P.J. Knowles, H.J. Werner, Chem. Phys. Lett. **115**(3), 259 (1985)
56. H.J. Werner, P.J. Knowles, J. Chem. Phys. **89**(9), 5803 (1988)
57. P.J. Knowles, H.J. Werner, Chem. Phys. Lett. **145**(6), 514 (1988)
58. H. Lischka, R. Shepard, I. Shavitt, R.M. Pitzer, M. Dallos, T. Müller, P.G. Szalay, F.B. Brown, R. Ahlrichs, H.J. Böhm, A. Chang, D.C. Comeau, R. Gdanitz, H. Dachsel, C. Ehrhardt, M. Ernzerhof, P. Höchtl, S. Irle, G. Kedziora, T. Kovar, V. Parasuk, M.J.M. Pepper, P. Scharf, H. Schiffer, M. Schindler, M. Schüler, M. Seth, E.A. Stahlberg, J.G. Zhao, S. Yabushita, Z. Zhang, M. Barbatti, S. Matsika, M. Schuurmann, D.R. Yarkony, S.R. Brozell, E.V. Beck, J.P. Blaudeau, M. Ruckenbauer, B. Sellner, F. Plasser, J.J. Szymczak, COLUMBUS, an ab initio electronic structure program, Release 7.0 (2012)
59. N. Johnson, O. Herrwerth, A. Wirth, S. De, I. Ben-Itzhak, M. Lezius, B. Bergues, M. Kling, A. Senftleben, C. Schröter, R. Moshammer, J. Ullrich, K. Betsch, R. Jones, A. Sayler, T. Rathje, K. Rühle, W. Müller, G. Paulus, Phys. Rev. A **83**(1), 013412 (2011)
60. T. Wittmann, B. Horvath, W. Helml, M. Schätzel, X. Gu, A. Cavalieri, G. Paulus, R. Kienberger, Nat. Phys. **5**(5), 357 (2009)
61. A.M. Sayler, T. Rathje, W. Müller, C. Kürbis, K. Rühle, G. Stibenz, G.G. Paulus, Opt. Express **19**(5), 4464 (2011)
62. A.M. Sayler, T. Rathje, W. Müller, K. Rühle, R. Kienberger, G.G. Paulus, Opt. Lett. **36**(1), 1 (2011)
63. M. Lewenstein, K. Kulander, K. Schafer, P. Bucksbaum, Phys. Rev. A **51**(2), 1495 (1995)
64. S. Chelkowski, A. Bandrauk, A. Apolonski, Phys. Rev. A **70**(1), 013815 (2004)
65. S. Chelkowski, A. Bandrauk, Phys. Rev. A **71**(5), 053815 (2005)
66. H. Hasegawa, A. Hishikawa, K. Yamanouchi, Chem. Phys. Lett. **349**(1–2), 57 (2001)
67. H. Xu, T. Okino, K. Nakai, K. Yamanouchi, S. Roither, X. Xie, D. Kartashov, M. Schöffler, A. Baltuska, M. Kitzler, Chem. Phys. Lett. **484**(4–6), 119 (2010)

68. C. Cornaggia, P. Hering, Phys. Rev. A **62**(2), 023403 (2000)
69. G.G. Paulus, W. Becker, W. Nicklich, H. Walther, J. Phys. B: At. Mol. Opt. Phys. **27**(21), L703 (1994)
70. R. Kopold, W. Becker, H. Rottke, W. Sandner, Phys. Rev. Lett. **85**(18), 3781 (2000)
71. B. Feuerstein, R. Moshammer, D. Fischer, A. Dorn, C. Schröter, J. Deipenwisch, J. Crespo, Lopez-Urrutia, C. Höhr, P. Neumayer, J. Ullrich, H. Rottke, C. Trump, M. Wittmann, G. Korn, W. Sandner, Phys. Rev. Lett. **87**(4), 043003 (2001)
72. G. Yudin, M. Ivanov, Phys. Rev. A **64**(1), 013409 (2001)
73. F. Rosca-Pruna, M. Vrakking, Phys. Rev. Lett. **87**(15), 153902 (2001)
74. H. Stapelfeldt, T. Seideman, Rev. Mod. Phys. **75**(2), 543 (2003)
75. T. Seideman, E. Hamilton, Adv. At. Mol. Opt. Phys. **52**, 289 (2005)
76. A. Alnaser, X. Tong, T. Osipov, S. Voss, C. Maharjan, B. Shan, Z. Chang, C. Cocke, Phys. Rev. A **70**(2), 23413 (2004)
77. X. Andrade, J. Alberdi-Rodriguez, D.A. Strubbe, M.J.T. Oliveira, F. Nogueira, A. Castro, J. Muguerza, A. Arruabarrena, S.G. Louie, A. Aspuru-Guzik, A. Rubio, M.A.L. Marques, J. Phys. Condens. Matter: Inst. Phys. J. **24**(23), 233202 (2012)
78. A. Castro, H. Appel, M. Oliveira, C.a. Rozzi, X. Andrade, F. Lorenzen, M.A.L. Marques, E.K.U. Gross, A. Rubio, Phys. Status Solidi B **243**(11), 2465 (2006)
79. M. Marques, A. Castro, G.F. Bertsch, A. Rubio, Comput. Phys. Commun. **151**(1), 60 (2003)
80. B. Friedrich, D. Herschbach, Phys. Rev. Lett. **74**(23), 4623 (1995)
81. R. Thissen, J. Delwiche, J.M. Robbe, D. Duflot, J.P. Flament, J.H.D. Eland, J. Chem. Phys. **99**(9), 6590 (1993)
82. T. Osipov, T.N. Rescigno, T. Weber, S. Miyabe, T. Jahnke, A.S. Alnaser, M.P. Hertlein, O. Jagutzki, L.P.H. Schmidt, M. Schöffler, L. Foucar, S. Schössler, T. Havermeier, M. Odenweller, S. Voss, B. Feinberg, A.L. Landers, M.H. Prior, R. Dörner, C.L. Cocke, A. Belkacem, J. Phys. B: At. Mol. Opt. Phys. **41**(9), 091001 (2008)
83. S. Bubin, M. Atkinson, K. Varga, X. Xie, S. Roither, D. Kartashov, A. Baltuška, M. Kitzler, Phys. Rev. A **86**(4), 043407 (2012)
84. E. Lötstedt, T. Kato, K. Yamanouchi, Phys. Rev. A **85**(4), 041402 (2012)
85. E. Lötstedt, T. Kato, K. Yamanouchi, Phys. Rev. A **86**(2), 023401 (2012)
86. F. Légaré, I. Litvinyuk, P. Dooley, F. Quéré, A. Bandrauk, D. Villeneuve, P. Corkum, Phys. Rev. Lett. **91**(9), 093002 (2003)

Chapter 4
Optimal Pulse Shaping for Ultrafast Laser Interaction with Quantum Systems

Hyosub Kim, Hangyeol Lee, Jongseok Lim and Jaewook Ahn

Abstract Coherent control method steers a quantum system to a desirable final quantum state among a number of final states otherwise possible in a given light-matter interaction, by using a specially shaped light form programmed in its spectral and/or temporal domain. In this chapter, we briefly review a number of light-form shaping methods previously considered for coherent control of ultra-fast laser interaction with atoms, and provide their application examples along with their experimental demonstrations.

4.1 Introduction

Latest development in laser optics manifests a new capability of light, other than the traditional use of light as a viewing tool, which is the use of light as a control tool. In particular, along with the advent of ultra-fast optical techniques, it becomes possible that the versatile light forms are newly engineered by programming the amplitude and shape functions of the broad frequency range of an ultra-fast optical pulse. As a result, the programmed spectral and/or temporal shape of a laser pulse can be used as a quantum-mechanical means to control the dynamics of a quantum system. This field of optical research is refereed to as coherent control, or quantum control, or more specifically to emphasize the usage of ultra-fast laser, "ultra-fast quantum control."

Examples of the coherent control concepts demonstrated in terms of shaped light-matter interactions can be found in numerous experiments. To list a few, historically Wilson and coworkers [1] experimentally realized closed-loop feedback control of optimal pulse shape for efficient population transfer in molecular system. Gerber

---

H. Kim · H. Lee · J. Lim · J. Ahn (✉)
Department of Physics, Korea Advanced Institute of Science and Technology (KAIST),
291 Daehak-ro, Yuseong-gu, Daejeon 305-701, Korea
e-mail: jwahn@kaist.ac.kr

K. Yamanouchi et al. (eds.), *Progress in Ultrafast Intense Laser Science XI*,
Springer Series in Chemical Physics 109, DOI: 10.1007/978-3-319-06731-5_4,
© Springer International Publishing Switzerland 2015

and coworkers [2] also demonstrated the coherent control in the photo-dissociation process of molecules, by using shaped ultra-fast optical pulses which were adaptively programmed in a closed control loop consisted of a laser programming apparatus and a photo-fragment mass spectroscope. Although the adaptive control methods are powerful in many real applications, we leave this class of coherent control methods aside and focus on the open-loop coherent control method, where the laser pulse shape is first designed by solving the Schrödinger equation for the given light-matter interaction. In that context, Silberberg and coworkers [3] initiated the pulse shape design research in terms of coherent control in their demonstration of atomic nonlinear absorption process. Other examples can be found, for example, in optical second-harmonic generation [4], optical third-harmonic generation [5], multi-photon absorption [6, 7], and coherent anti-Stokes Raman scattering [8]. More advanced applications can be found in in-vivo fluorescence microscopy [9], coherent control quantum bits in Rydberg atoms [10], and semiconductor [11] to list a few.

The research of coherent control of quantum dynamics has become possible in many ways thanks to the development of laser pulse shaping apparatus in the last two decades [12]. Among many apparatus, spatial light modulators (SLM), in particular, allow a relative easy way to engineer the interference among the transition pathways of a multi-photon absorption process, because of their Fourier domain spectral shaping capability [3, 6]. Also, acousto-optic programmable dispersive filters (AOPDF) provide more sophisticated ways of pulse-shaping in the context of this chapter, showing its capability, for example, in the strong-field multi-photon absorption control of alkali atoms [7, 13]. A combination of those pulse-shaping devices can transform a well-defined gaussian-shape pulse to an arbitrary pulse shape of programmed spectral phase, amplitude, and also polarization [14–18].

4.2 Types of Pulse Shaping

Before we begin to describe the coherent control experiments, we categorize the ways of pulse shaping methods. To do that, we first need to define an ultra-fast laser pulse.

The ultra-fast laser pulse can be represented either in time domain as $E(t)$, or in frequency domain $\tilde{E}(\omega)$, where their mutual relation is given by the complex Fourier transform ($\mathscr{F}$). The most interesting case in this context of ultra-fast optical pulse shaping is a moderately short pulse, which means the center carrier frequency ω_0, and the bandwidth of the pulse $\Delta\omega$ are related as $\Delta\omega \ll \omega_0$ [19]. Then, without a loss of generality, the electric field is factored into the envelope and carrier functions given by

$$E(t) = \frac{1}{2}\varepsilon(t)e^{i\phi(t)}e^{i\omega_0 t} + c.c., \tag{4.1}$$

4 Optimal Pulse Shaping for Ultrafast Laser Interaction

and its frequency domain representation is given by

$$\tilde{E}(\omega) = \mathscr{F}\{E(t)\} = |\tilde{E}(\omega)|e^{i\Phi(\omega)}, \tag{4.2}$$

where the tilde denotes a complex value and, since $E(t)$ is real function, a relation $\tilde{E}(\omega) = \tilde{E}^*(-\omega)$ holds.

It is noted that we neglect the spatial profile of the laser pulse, which often plays an important role in nonlinear optical process in terms of spatial average effect [20], but we assume the given light-matter interaction is performed in an uniform spatial intensity region only.

As seen in (4.1), the electric field of the pulse is described by its center carrier frequency ω_0 and its transient envelope function $\varepsilon(t)$. The Fourier transform of this temporal profile yields the same analogy in the frequency domain, which is centered at ω_0. Here, we call $|\tilde{E}(\omega)|$ as a spectral amplitude and $\Phi(\omega - \omega_0)$ as a spectral phase. It is noted that the spectral phase is a function of frequency, and, in particular, if the spectral phase is a constant over the whole spectrum range, the laser pulse is referred to as a Fourier transform-limited pulse.

The spectral phase can be expanded into a Taylor series as

$$\Phi(\omega) = \Phi(\omega_0) + \Phi'(\omega - \omega_0) + \frac{1}{2}\Phi''(\omega - \omega_0)^2 + \frac{1}{6}\Phi'''(\omega - \omega_0)^3 + \cdots, \tag{4.3}$$

where the constant phase $\Phi(\omega_0)$ is the carrier-envelop phase, Φ' is the movement of pulse envelope along the positive time direction, (which is in fact equivalent to the frequency shift of ω_0), Φ'' is the linear spectral chirp, and Φ''' is the quadratic spectral chirp, etc. As a pulse shape contains both the envelop shape and phase information at the same time, pulse shaping control is equivalent to both the amplitude and phase shape manipulation of the laser pulse. Also, as the actually time and frequency domains are related with respect to the Fourier transform relation, only one of the domain control, either time or frequency, is sufficient for the generation of a desired shape programming of a laser pulse.

The pulse shaping device generally operates in spectral domain, to be discussed in Sect. 4.3, so we can categorize the types of pulse shaping in spectral domain as following:

1. Amplitude shaping (lose photons: irreversible shaping)

   a. Spectral amplitude blocking
   b. Wavelength scanning (regular spectroscopy)

2. Phase shaping (temporal rearrangement of spectral components)

   a. Spectral chirping
   b. Multiple pulses (e.g., 2D-Fourier transform spectroscopy)
   c. Spectral phase gating
   d. Spectral phase step

The amplitude shaping methods in spectral domain inevitably lose photons, and, therefore, this type of pulse shaping methods is intrinsically an irreversible process. However, to be described in Sect. 4.4, certain light-matter interactions are better performed with a smaller set of spectral components of light. For example, in the two-photon absorption case in a three-level atom with an intermediate energy level, a certain part of laser spectrum participates to the net absorption as a destructive quantum interference. Therefore, removing those spectral components, to make the remaining absorption pathways all constructively interfere with each other, could enhance the net absorption probability. This method is known as spectral blocking [6, 21, 22]. Another method is a wavelength scanning which could be implemented by allowing only a narrow spectral region of an ultra-fast pulse and blocking the others.

The spectral phase shaping is a way of temporal rearrangement of spectral components, and this method allows a variety of ways of spectral manipulation of a laser pulse. For example, spectral phase gating method provides a narrow spectral window region (phase-gated window) to be of different phase from the others, and scans the phase-gated spectral window from one end to the other end of the laser spectrum. Also, one can make a spectral phase step in a certain spectral location of the spectrum and scan the phase-step location, or one can implement a modulated spectral phase function $\Phi(\omega) = A\cos(B\omega + C)$ and by controlling the modulation amplitude A, the modulation frequency B, or the modulation phase C, respectively, the population transfer among the energy levels of the given light-matter interaction can be enhanced or suppressed. Another example is a use of multiple laser pulses, which can be programmed in the phase-shaping in spectral domain. For example, a set of three laser pulses with a fixed carrier-envelope phase and variable time delays implements two-dimensional Fourier transform spectroscopy. One could change the frequency de-tuning, the spectral linear chirp, or the spectral quadratic chirp by changing Φ', Φ'', or Φ''', respectively, which subject will be discussed in more detail in Sect. 4.5. Lastly, if the transition phase information is completely known a priori, then a spectral phase function which maximizes the given transition can be programmed to be discussed in Sect. 4.6.

4.3 Pulse Shaping Devices

Although the pulse shaping methods are performed in both the Fourier and time domains, the tailored pulse can be described in either domain completely. Thus we concentrate on the frequency domain[1] shaping here. When we deal with a pulse at frequency domain, angular dispersive optical elements, such as a prism or a diffraction grating, can be used [23, 24]. As such optical elements disperse each chromatic component at a different spatial position in the real plane, managing each chromatic

---

[1] We notice that Fourier domain and frequency domain have the same physical meaning in this chapter.

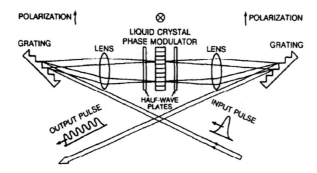

Fig. 4.1 An optical pulse shaper combined with a pair of diffraction gratings and a liquid crystal SLM. Phase, and amplitude, or both phase and amplitude are controllable according to respective types of configurations of an SLM, inserted wave plates, and polarizers. The image is from [27] with permission by the publisher

ray at different position would possibly manipulate the shape of the pulse in the frequency domain. In the contrary to frequency domain, time domain control can be done by acousto-optic effect [16]. Programmable acoustic waves generated by piezo-electric transducer in an acousto-optic crystal convolute with incident pulses and generate desired pulse shape.

Spectral amplitude and phase functions can be controlled either by a fixed mask [23], or by a variable masks: For example, SLM [25, 26], AOPDF [17], and etc [27]. We briefly describe the working principles of SLM and AOPDF, before we start to discuss the experiments performed with these devices.

4.3.1 Spatial Light Modulator

SLM is an array of optical phase or amplitude (or both) modulators made out of liquid crystals. The liquid crystal in a nematic phase is a uniaxial birefringent material and its molecular orientation is easily controllable by means of an applied electric field. A usual setup, shown in the Fig. 4.1, is used as an optical pulse shaper, where an SLM pulse shaper is located in the Fourier plane of a 2f-to-2f configuration of lens and diffraction grating setup, also known as Martinez zero-dispersion stretcher setup [27, 28]. The phase of light transmitted through each liquid crystal pixel is given as a function of both the incident polarization and the electric field applied to the pixel. For example, in the case of a phase-only modulation SLM, the extra-ordinary axis of the liquid crystal is oriented originally parallel to incident polarization and, when the electric field is applied to the propagating direction of light, optical path length can be managed without birefringence. With appropriate combinations of half-wave plates and polarizers, amplitude-modulation and both amplitude-and-phase modulation are also possible.

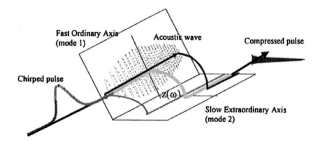

Fig. 4.2 Schematic of the AOPDF. Traveling acoustic wave diffracts phase-matched optical mode. The image is from [17] with permission by the publisher

4.3.2 Acousto-Optic Programmable Dispersive Filter

The working principle of AOPDF is based on collinear acousto-optic interaction, where a programmed acoustic wave diffracts different spectral components, of an ultra-fast laser pulse, that are respectively phase-matched with different acoustic modes in the AOPDF [16]. Thus, the output laser spectral amplitude E_{out} is programmed as $E_{\text{out}} \propto E_{\text{in}} S(\alpha\omega)$, where E_{in} is the input laser spectral amplitude and $S(\alpha\omega)$ is the acoustic mode amplitude and α is the ratio between the speed of sound and the speed of light in the crystal (Fig. 4.2).

4.4 Spectral Amplitude Blocking

4.4.1 A Ladder-Type System

As the first example of coherent control experiments, we consider the spectral blocking among spectral amplitude shaping methods. The quantum system under consideration is the three-level system in a ladder configuration of atomic rubidium (Rb). The three energy levels are the 5S (the ground state), $5P_{1/2}$ (the intermediate state), and 5D states (the final state), as shown in the Fig. 4.3. The coherent light source is from a mode-locked Ti:sapphire laser, of which the wavelength center is at 778 nm and the broad wavelength width of 18 nm in FWHM (the full width at half maximum) covers both the upper and lower intermediate states energy separation.

The transition probability amplitude to the final state is given by the second order time dependent perturbation theory [6] as,

$$c^{(2)}(t) = -\frac{1}{\hbar^2} \sum_i \mu_{fi}\mu_{ig} \int_{-\infty}^{t} \int_{-\infty}^{t_1} E(t_1)E(t_2)e^{i\omega_{fi}t_1}e^{i\omega_{ig}t_2}dt_2 dt_1, \quad (4.4)$$

Fig. 4.3 Energy level diagram of the ladder-type two-photon absorption

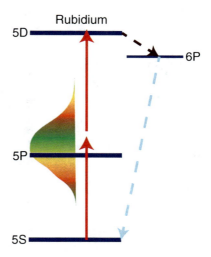

where μ_{fi} and μ_{ig} are the transition dipole moments. The angular frequency is defined as energy spacing between the two levels as $\omega_{ij} \equiv (E_i - E_j)/\hbar$. We consider an electric field $E(t)$ given as a pulse as,

$$E(t) = \frac{1}{\sqrt{2\pi}} \int_{-\infty}^{\infty} \tilde{E}(\omega) e^{-i\omega t} d\omega. \tag{4.5}$$

Now, the both $E(t_1)$ and $E(t_2)$ are substituted into (4.4) and the complex integral is conducted with an assumption that fast transient is neglected, or the integration time t is large enough compared to the pulse duration. When the spectral bandwidth covers both the transitions ω_{fi} and ω_{ig}, respectively, from the ground state to the intermediate state and from the intermediate state to the final state, the solution is given by

$$c_{fg}^{(2)} = -\frac{1}{i\hbar^2} \mu_{fi} \mu_{ig} [i\pi \tilde{E}(\omega_{ig}) \tilde{E}(\omega_{fg} - \omega_{ig}) + \mathcal{P} \int_{-\infty}^{\infty} \frac{\tilde{E}(\omega) \tilde{E}(\omega_{fg} - \omega)}{\omega_{ig} - \omega} d\omega], \tag{4.6}$$

where the first term corresponds to the resonant two-photon transition and the second term, the Cauchy integral term, exhibits the broad non-resonant behavior of the two-photon absorption. Such a broad spectral response is originated from the short temporal mediation of the intermediate state and thus a large uncertainty in energy occurs [6], but, as the excited state has a rather long lifetime, the total energy of two photons is conserved. It is noted that the transition probability amplitude of the resonant and non-resonant components have different phase. Especially the constituent

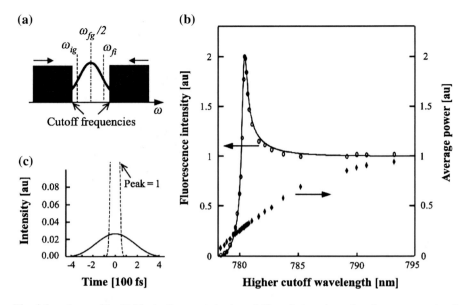

Fig. 4.4 a A movable slit blocks the spectral wings. **b** Two photon absorption fluorescence signal with respect to the cutoff wavelength and the corresponding average power. **c** Comparison between the transform limit pulse and the optimized two-photon absorption pulse. The image is from [6] with permission by the publisher

contributions in the non-resonant term destructively interfere within themselves. Therefore, definitely, a transform-limit pulse is not optimal for the given two-photon absorption. Figure 4.4 shows experimental demonstration that the spectral amplitude blocking of the spectral component dramatically increases transition probability amplitude while the pulse energy significantly decreases. In the described experiment, the spectral blocking experiment was carried out by a mechanical slit in the Fourier domain.

4.4.2 Spectral Amplitude Blocking in a V-Type System

The spectral blocking method can be also used for a V-type three level system. To summarize the working principle, the experimental example is explained below.

As shown in Fig. 4.5, the system under consideration is the three energy levels of atomic rubidium, $|g\rangle = 5S$, $|a\rangle = 5P_{1/2}$, and $|b\rangle = 5P_{3/2}$. The two fine structure levels $|a\rangle$ and $|b\rangle$ are simultaneously excited by an ultra-fast laser pulse from the common ground state $|g\rangle$ [21]. The particular transition that we try to control is the two-photon transition from $|a\rangle$ to $|b\rangle$ via $|g\rangle$, the V-type two-photon transition. The corresponding transition probability amplitude can be calculated in a similar manner to the ladder type case, and the resulting formula as a result of the second

4 Optimal Pulse Shaping for Ultrafast Laser Interaction

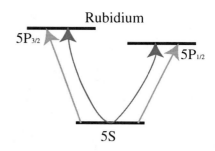

Fig. 4.5 The V-type system of Rb atom. Two-adjacent levels are forbidden transitions

order time dependent perturbation calculation is given by [22],

$$c_{ba}^{(2)} = \frac{1}{i\hbar^2}\mu_{ga}\mu_{gb}[-i\pi \tilde{E}^*(\omega_{ag})\tilde{E}(\omega_{bg}) + \mathscr{P}\int_{-\infty}^{\infty}\frac{\tilde{E}^*(\omega)\tilde{E}(\omega_{ba}+\omega)}{\omega_{ag}-\omega}d\omega], \quad (4.7)$$

where it is noted that the one electric field is the conjugate pair compared to the result in (4.6) of which the feature is originated from the down and upward sequence of the participating three energy states transition.

The experimental observation of the V-type transition to $|b\rangle$ from $|a\rangle$ via $|g\rangle$, which is carried out by detecting the population of $|b\rangle$, needs to be differentiated from the simple one photon transition to $|b\rangle$ from $|g\rangle$. Therefore, the experiments were performed by using a two-dimension Fourier transform spectroscopy technique. By using three optical pulses (the initial preparation pulse, the control pulse, and the final interference pulse, respectively) with different phase oscillation between each states. At the same time, the control pulse was spectrally shaped to implement the spectral blocking for the V-type transition by using AOPDF.

Figure 4.6 shows the experimental results of the enhancement of the two-photon absorption in the V-type system. As evident from the result, the shaped pulse, which has lesser energy than a transform-limited pulse, demonstrates the enhancement of the two-photon transition probability amplitude in the given atomic systems. This results contradicts our intuition that the given non-linear transition is maximized by a transform-limited pulse, which has the maximal peak intensity. However, as the theoretical formula predicts, the given broad-band two-photon transition should be enhanced by eliminating the destructive interference contributions. In the same line of thought, then one could come up with a more efficient way to increase the transition not only by eliminating the destructive interference contributions but by controlling them to constructively interfere with the rest of the transition components, in particular by spectral phase shaping, to be explained in Sect. 4.6.1

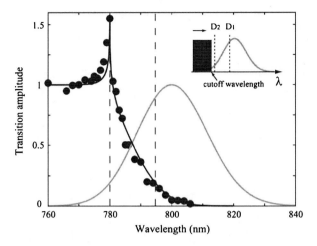

Fig. 4.6 5P$_{1/2}$-5P$_{3/2}$ two-photon transition probability amplitude controlled by the spectral amplitude blocking method. *Dots* represent the experimental measurement and the *dark line* the theoretical calculation. The *grey line* shows the spectrum of the laser pulse. Inset shows the way how the spectrum is blocked. The image is from [21] with permission by the publisher

4.5 Spectral Chirp Control

Chirping is also a crucial parameter for manipulating coherent population control in atom-light interaction. Chirping and quadratic chirping in the frequency domain are described in (4.3) as a Taylor expansion of spectral phase near the center frequency. However, the physical meaning of the linear chirp, or linear spectral chirp, is rather simply understood at the time domain; the oscillating period ω_0 changes as a time dependent variable as $\omega_0 + \beta t$, where the chirp parameter is defined by $\beta = 2\Phi''/(\tau_0^4 + 4\Phi''^2)$ with the time duration τ_0 defined for a corresponding transform-limited pulse (field).[2]

4.5.1 Chirps in a 2 + 1 Photon Transition

The first example of chirp-control experiment is the 2 + 1 photon transition in an asymmetric three-level ladder system of atomic sodium as shown in Fig. 4.7 [7]. The three-level energy ladder is consisted of the 3S, 4S, and 7P energy states, and the first 3S–4S transition is a two-photon transition, and the second 4S–7P transition is a one-photon transition.

The laser pulses from the femto-second laser amplifier operating at a repetition rate of 1 kHz have the transform-limited pulse duration of 37 fs. Within the allowed

---

[2] (field) means that the description is about electric field. Readers should care whether (field) or (intensity) arise at the end of a description.

4 Optimal Pulse Shaping for Ultrafast Laser Interaction

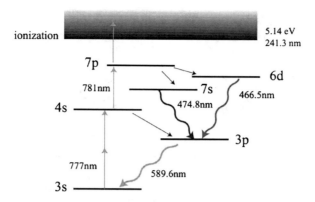

Fig. 4.7 The asymmetric three-level energy ladder in atomic sodium and the 2+photon transition pathway from 3S to 7P via 4S. The direct four-photon ionized atom signal is 1,000 times smaller than the 4S-7P sequential 2 + 1 photon transition, thus the system is valid as a closed three-level system. The image is from [7] with permission by the publisher

laser bandwidth there is no intermediate state during the two-photon 3S–4S transition, and, therefore, the 3S–4S transition is non-resonant.

When the chirp rate was controlled by AOPDF, as in Fig. 4.8, the 7P state population was measured as a function of the chirp rate of the excitation laser pulse. The experimental results show highly asymmetric behavior of the three photon excited state population given as a function of the chirp rate. The physical origin of the asymmetry can be explained by the different excitation pathways: One is the 4S-state mediated sequential 2 + 1 photon excitation path and the other is a direct three photon excitation path. In the case of the former path, the first two-photon excitation has a higher frequency (777 nm) and the last one-photon excitation has a lower frequency (781 nm), thus the transient character of the excitation process prefers a negative chirp rate and, therefore, is the main reason of the observed asymmetry. The latter path has no reason to have such an asymmetry, rather it shows symmetric behavior with zero population at zero chirp rate, which can be easily observed in the green dot-dashed line in Fig. 4.8. The reason can be understood based on the effect of the dynamic Stark shift which makes the intensity proportional resonance level de-tuning in the presence of the laser pulse [29]. In this scenario, the zero chirp point has the highest intensity, because the largest dynamic Stark shift occurs, and the direct three photon transition path becomes far off de-tuned from the resonance energy condition, which explains the symmetric nature of the intensity shape with a dip at the zero chirp region. Interference between direct path and sequential path causes also considerable effect in multi-photon transition system [30]. However, in the 2 + 1 photon system, calculation based on perturbation theory shows that chirp rates in the experiment are not enough to observe interference fringes.

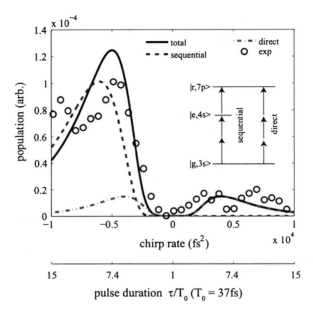

Fig. 4.8 Sodium 2 + 1 photon transition measured as a function of the chirp rate. Experimental results plotted by dots are compared with the theoretical calculation (*the lines*). The 2 + 1, photon sequential transition pathway is in the *blue dashed line* and the three photon direct transition is in the *green dot-dashed line* which are asymmetric and symmetric respectively. The *black solid line* represents the total absorption rate. The image is from [7] with permission by the publisher

4.5.2 Chirps in Two-Photon Transitions

The excitation suppression due to the dynamic Stark effect can be better identified by observing the 3S−4S two-photon transition. As shown in Fig. 4.9, 3S–4S two-photon direct transition path is dominant compared to the four-photon sequential transition path,[3] and, therefore, an observation of the 4s state population as a function of the chirp rate provides a clear evidence for the dynamic Stark effect induced transition reduction.

Experimental result in atomic cesium also finds a similar effect. As a two-photon transition without any resonant intermediate transition, which often referred to as a non-resonant two-photon transition, the 6S–8S transition in cesium is strongly influenced by the dynamic Stark shift in the presence of an intense laser field. In the experiment carried out with a laser intensity around 10^{12} W/cm^2, or in the strong field regime, the chirp rate dependence on the two-photon transition has been investigated [13].

Theoretical consideration for the $3S − 4S$ two-photon transition probability gives an analytical formula given as a function of the chirp rate and the pulse intensity, for

---

[3] Four-photon sequential path is 3S−4S − 7P 2 + 1 photon process in addition to 7P−4S one-photon de-excitation. In this experiment, several order of magnitude is smaller than two-photon direct path.

4 Optimal Pulse Shaping for Ultrafast Laser Interaction

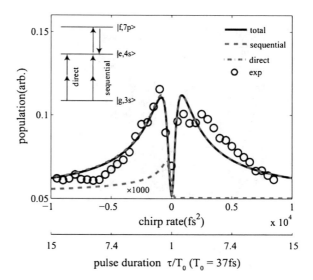

Fig. 4.9 Sodium two-photon absorption rate measured as a function of the chirp rate. The two-photon direct transition path turns out much larger than the four photon sequential transition path. The image is from [7] with permission by the publisher

the case of a zero de-tuning from the two-photon resonant condition, by

$$P_e \propto \frac{1}{\sqrt{1+4a_2^2/\tau_0^4}} \exp[-\frac{\pi \eta^2 I_0^2 \tau_0^2}{8(1+4a_2^2/\tau_0^4)}], \quad (4.8)$$

where a_2, τ_0, and I_0 are the chirp rate,[4] transform-limited pulse width (intensity) defined by $\exp(-t^2/\tau_0^2)$, and the peak intensity respectively. η in the equation has complicated physical origin, thus we just treat it as a fitting constant here.[5] Experimental result obtained as a function of the chirp rate and the pulse intensity is shown in Fig. 4.10. As the pulse intensity increases, a single peak centered at the zero chirp region becomes divided into double peaks with a sharp dip in the zero chirp, which behavior is similar to the case of Fig. 4.9. The exponential decay term in (4.8) properly describes such a behavior: As the pulse intensity is big enough so that the dynamic Stark shift plays a role, the signal given as a function of the chirp rate appears with a dip point around the zero chirp with a symmetric double side peaks as shown in Fig. 4.10b. However, when the pulse intensity is low, the signal shows a usual behavior of a single peak around the zero chirp, as shown in Fig. 4.10c. The contour lines in Fig. 4.10 are theoretically calculated from (4.8).

---

[4] The linear spectral chirp Φ'' in (4.3) and a_2 here is identical parameter.

[5] For more detailed description, we would recommend readers to read [13].

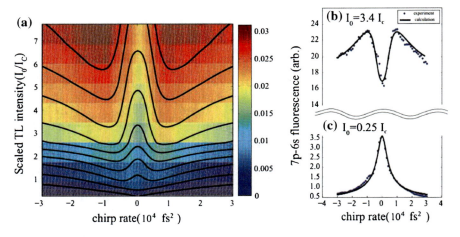

Fig. 4.10 **a** Two-photon transition probability to cesium 8 s state from the ground 6 s state measured as a function of chirp rate and pulse intensity. **b** and **c** show the behaviors in the strong and weak interaction regimes, respectively. The image is from [13] with permission by the publisher

So far, we have seen that the effect of dynamic Stark shift can decrease the two-photon transition in the strong field regime. The chirp rate, both negative and positive, could increase the transition rate by reducing the amount of energy level shift.

4.5.3 Optimal Pulse Shaping of a Two-Photon Transition

Then, one can also consider a case that a more sophisticated pulse shaping method could even compensate for the dynamic energy level shift to increase the transition [31]. The idea has been experimentally realized in the Cesium 6S–8S non-resonant two-photon transition case [32].

The cesium 6S–8S non-resonant two-photon transition can be described by a two-level system Hamiltonian [32],

$$H(t) = \hbar \begin{bmatrix} 0 & \frac{1}{2}\Omega(t)e^{2i\phi(t)} \\ \frac{1}{2}\Omega(t)e^{-2i\phi(t)} & \Delta + \delta(t) \end{bmatrix}, \quad (4.9)$$

by appying the unitary transformation $U(t) = |g><g| + e^{-2i\phi(t)}|e><e|$,

$$H(t) = \hbar \begin{bmatrix} 0 & \frac{1}{2}\Omega(t) \\ \frac{1}{2}\Omega(t) & \Delta + \delta(t) - 2\dot{\phi}(t) \end{bmatrix}, \quad (4.10)$$

4 Optimal Pulse Shaping for Ultrafast Laser Interaction

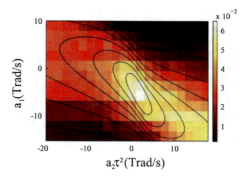

Fig. 4.11 Two-photon transition rate between the cesium 6S and 8S energy levels given as a function of the two control parameters a_1 and a_2 defined in (4.11). Measurements were carried out at the laser intensity of 0.21×10^{11} W/cm^2. The image is from [32] with permission by the publisher

where $\Omega(t)$ is the two-photon Rabi frequency and $\phi(t)$ is the phase of electric field. $\Delta = \omega_e - \omega_g - 2\omega_0$ is the de-tuning and $\delta(t)$ is the effective dynamic Stark shift. The $\delta(t)$ is proportional to intensity $I(t)$.

It is known that the population transfer by the Hamiltonian, (4.10) can be maximized by making the diagonal term of the Hamiltonian zero [31]. For this, the temporal pulse phase should satisfy the condition $2\dot{\phi}(t) = \Delta + \delta(t)$, or the system maintains dynamically the resonant condition. To compensate the dynamic Stark shift, $\delta(t)$, which is proportional to the temporal pulse intensity $\exp(-t^2/\tau^2)$, we expand $\delta(t)$ as a Taylor series up to second term. Then, the compensation condition is given by,

$$2\phi(t) = (\Delta + \delta(0))t - \frac{1}{3}\frac{\delta(0)}{\tau^2}t^3 = 2a_1 t + 2a_2 t^3 \qquad (4.11)$$

The temporal phase is described by a polynomial form similar to the (4.3) of the spectral phase. The zeroth order and the second order terms of the temporal phase have the same physical origins, respectively, as those of spectral phase, which are carrier envelop phase and chirp rate, respectively. While, the first order term of temporal phase, on the other hand, is different from that of spectral phase. The former corresponds to a shift of the frequency envelope and the latter the time envelope. The third order term of the temporal phase is hardly described by the spectral phase components only, because it needs both controls of the spectral amplitude and phase at the same time.

The experimental result in Fig. 4.11 shows a good agreement with the calculation that is depicted by the black contour lines. The calculation is the numerical result of the time dependent shröedinger equation based on the Hamiltonian in (4.10). The approximation terms of the dynamic Stark shift $\delta(t)$ up to second order is used. The intensity are kept low enough to prevent loss through the ionization passage. The parameter a_1 shifts the two-photon energy of the laser pulse toward the shifted 8S energy level near the resonant condition, and a_2 bends the temporal frequency shape to have a more overlapped temporal region with the temporal shape of the dynamic Stark shift. Higher order terms neglected in (4.11) would better supplement the compensation if considered.

4.5.4 Chirps in a V-Type System

It is worth to mention the effects of chirps on the V-type two-photon transition [22]. The transition probability of (4.7) for the V-type two-photon transition can be simplified when the pulse envelope is assumed to be of a gaussian shape, and the spectral phase has a chirp (a_2) and a quadratic chirp (a_3) only. After a straightforward algebra, the given 2-photon transition can be written as

$$c_{ba}^{(2)} = i \frac{\tilde{\mu}_{ba}}{\hbar^2} [i\pi \tilde{\mathscr{E}}(\bar{\omega}) - \mathscr{P} \int_{\infty}^{\infty} \frac{\tilde{\mathscr{E}}(\omega)}{\bar{\omega} - \omega} d\omega], \tag{4.12}$$

where $\tilde{\mu}_{ba} = \mu_{ga}\mu_{bg} \exp[-\omega_{ba}^2/2\Delta\omega^2 + ia_3\omega_{ba}^3/24]$, $\bar{\omega} = (\omega_{ag} + \omega_{bg})/2$, and $\tilde{\mathscr{E}}(\omega) = |\tilde{E}(\omega)|^2 \exp(i\omega_{ba}d\phi/d\omega)$ is the effective electric field. We just write down the one-photon de-excitation probability amplitude calculated by first order perturbation [22],

$$c_{ge}^{(1)} = \frac{i\mu_{eg}E_0}{\hbar} \int_{-\infty}^{t} \exp(-\frac{t'^2}{\tau_c^2}) \times \exp\{-i[(\omega_{eg} - \omega_0)t' - \alpha t'^2]\}dt'$$

$$\underset{=}{F.T} \frac{\mu_{eg}}{\hbar} [i\pi \tilde{E}^*(\omega_{eg}) - \mathscr{P} \int_{-\infty}^{\infty} \frac{\tilde{E}^*(\omega)e^{i(\omega-\omega_{eg})t}}{\omega_{eg} - \omega} d\omega], \tag{4.13}$$

where $\tau_c = \tau_0\sqrt{1 + a_2^2/\tau_0^4}$, $\alpha = 2a_2/(\tau_0^4 + 4a_2^2)$, and τ_0 is the transform-limited pulse width (field) [22]. The readers should notice that the linear chirp rate is introduced in the de-excitation pulse for the case of (4.13). At time $t = 0$, (4.12) and the second (4.13) are of a similar form. Thus, the V-type two-photon transition process may be reduced to a one-photon de-excitation process. We conduct inverse Fourier transformation of (4.12) with an appropriate substitution of the effective electric field $\tilde{\mathscr{E}}(\omega)$ by a temporal shift $\tilde{t} = \omega_{ba}a_2$ and a chirp $\tilde{a}_2 = -\omega_{ba}a_3$. We can find its time domain form of (4.12) as

$$c_{ba}^{(2)} = -\frac{\tilde{\mu}_{ba}E_0^2 e^{i\theta}}{\hbar^2} \frac{\Delta\omega}{\sqrt{\tilde{\tau}_c/\tilde{\tau}_0}} \int_{-\infty}^{\tilde{t}} \exp(-\frac{t'^2}{\tilde{\tau}_c^2})$$

$$\times \exp\{-i[(\bar{\omega} - \omega_0)t' - \tilde{\alpha}t'^2]\}dt' \tag{4.14}$$

where $\theta = -\frac{1}{2}\tan^{-1} 2\tilde{a}_2/\tilde{\tau}_0^2 + (\bar{\omega} - \omega_0)\tilde{t}$, $\tilde{\tau}_0 = 2\sqrt{2}/\Delta\omega$, $\tilde{\tau}_c = \tilde{\tau}_0\sqrt{1 + \tilde{a}_2^2/\tilde{\tau}_0^4}$, and $\tilde{\alpha} = 2\tilde{a}_2/(\tilde{\tau}_0^4 + 4\tilde{a}_2^2)$.

We note that the V-type two-photon transition probability amplitude in (4.7) is a result of infinite temporal integration, but the reduced probability amplitude in (4.14)

4 Optimal Pulse Shaping for Ultrafast Laser Interaction

exhibits a time dependent-like form. The time dependence is in fact originated from the fact that the effective electric field is time-shifted proportional to the chirp rate, $\tilde{t} = \omega_{ba} a_2$. Interestingly, the V-type transition shows a coherent transient (CT)-like behavior as the chirp rate is controlled, although it should be distinguished from the real CT effect observed in pump-probe experiments [33–35].

Experimental result shown in Fig. 4.12I indeed exhibits the fore-mentioned CT-like behavior as a function of chirp rate. It is noted that three-pulse two-dimension Fourier transform spectroscopy can recover both the amplitude and phase of probability amplitude as shown in Fig. 4.12II.[6]

4.6 Spectral Phase Programming

One can also consider an arbitrary spectral phase function programming for the purpose of an optimized multi-photon transition. In particular, if the constituent transition pathways of the given transition are fully understood in terms of their respective phase information, maximally constructive quantum interference can be achieved by properly encoding the spectral phase function of the control laser pulse.

4.6.1 Spectral Phase Programming for a V-Type Transition

We consider the V-type two-photon system again, as an example. The target transition, between the adjacent fine structure energy levels through the ground state, is controlled by programming a spectral phase step at the spectral region $[\omega_{ag}, \omega_{ag} + \omega_{ba}]$. The governing equation is given in (4.7), where the integral term of (4.7) is decomposed into two part [21] as follows:

$$
\begin{aligned}
c_{ba}^{(2)} \simeq i \frac{\mu_{ga}\mu_{gb}}{\hbar^2} \Big[& i\pi \tilde{E}^*(\omega_{ag})\tilde{E}(\omega_{bg}) \\
& - e^{i\phi_b} \int_{\omega_{ag}-\omega_{ba}}^{\omega_{ag}} \frac{\tilde{E}^*(\omega)\tilde{E}(\omega_{ba}+\omega)}{|\omega_{ag}-\omega|} d\omega \\
& + e^{-i\phi_b} \int_{\omega_{ag}}^{\omega_{ag}+\omega_{ba}} \frac{\tilde{E}^*(\omega)\tilde{E}(\omega_{ba}+\omega)}{|\omega_{ag}-\omega|} d\omega \Big],
\end{aligned}
\tag{4.15}
$$

---

[6] Three ultra-fast pulses are incident on the target, with separately controllable pico-second scale time delays τ_1 and τ_2 between each consecutive pulse. The excited states induced by each pulse are interfered and the final fluorescence signal $I(\tau_1, \tau_2)$ is recorded. Two-dimensional Fourier transform of $I(\tau_1, \tau_2)$ acquires both amplitude and phase information. Detailed theoretical description on this experiment can be found from [22].

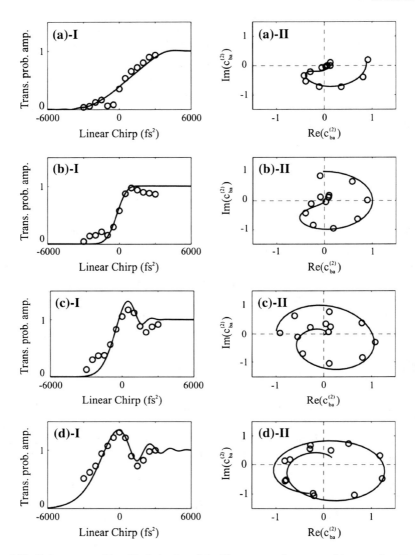

Fig. 4.12 Coherent transition-like behavior of the V-type two-photon transition as a function of linear chirp. The quadric chirps are given as **a** $-5 \times 10^4 fs^3$, **b** $-1 \times 10^4 fs^3$, **c** $3 \times 10^4 fs^3$, and **d** $7 \times 10^4 fs^3$. Sublabel **I** is the transition probability amplitude and **II** is the amplitude including the phase in complex plane. Solid lines are simulation and dots are experimental results. The image is from [22] with permission by the publisher

where the phase ϕ_b is the phase of the region $[\omega_{ag}, \omega_{ag} + \omega_{ba}]$, and the other regions have zero phase. The off-resonant components far from ω_{ag} are negligibly small thus omitted here.

The phase of the resonant part is fixed and the two non-resonant parts have rotating and counter-rotating phase, respectively. As the non-resonant components are

4 Optimal Pulse Shaping for Ultrafast Laser Interaction

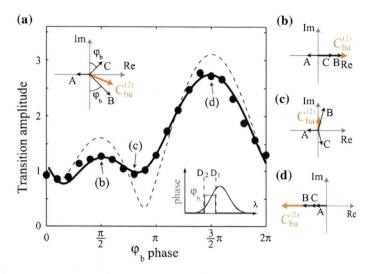

Fig. 4.13 **a** Transition amplitude with respect to the block spectral phase. Dots are experimental data, the *dashed line* is the simulation result based on (4.15), and the *solid line* is the simulation result with experimental systematic error. **b–d** Phase directions of each transition components of (4.15) with respect to the block spectral phase. The image is from [21] with permission by the publisher

initially, respectively, 90° out of phase from the resonant component, by manipulating the phase of the given spectral block, the transition amplitude shows an interference between the resonant and non-resonant components. The experimental and simulation results are shown in Fig. 4.13. The transition probability amplitude oscillates as the phase of the spectral block in $[\omega_{ag}, \omega_{ag} + \omega_{ba}]$ increases.

4.6.2 Spectral Phase Programming for a Non-resonant Two-Photon Transition

Another appreciable problem is the non-resonant two-photon transition [3]. As the non-resonant two-photon transition probability amplitude is written as a result of the second-order perturbation calculation as

$$c_{nr}^{(2)} \propto \int_{-\infty}^{\infty} |E(\omega_0/2 + \omega)E(\omega_0/2 - \omega)|e^{i[\phi(\omega_0/2+\omega)+\phi(\omega_0/2-\omega)]}d\omega, \qquad (4.16)$$

the transition has maximum when the spectral phase becomes antisymmetric ($\phi(\omega_0/2 + \omega) = -\phi(\omega_0/2 - \omega)$) or zero. However, it is more complicated when the spectral phase is symmetric. The transition probability can be zero or not. For the

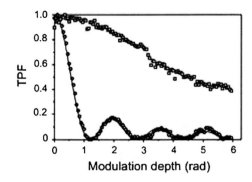

Fig. 4.14 Two-photon fluorescence with respect to modulation depth, A. The *Circle* and *square* data points are $C = 0$ and $C = \pi$ respectively. B is fixed at 220 fs. The image is from [3] with permission by the publisher

case of sinusoidal phase function, given by $A\cos(B\omega + C)$, the experiments has been performed with atomic cesium as in Fig. 4.14.

Then the two photon fluorescence signal measured as a function of the modulation depth A, for the cases of symmetric (circle) and antisymmetric (square) phase function. The antisymmetric sine phase predicts a flat, monotonic, two-photon transition, although it decays as the modulation depth grows. However, the result for a symmetric cosine phase shows much faster decay with periodic nodes, which behavior is well understood by the theoretical consideration in (4.14).

4.7 Conclusion

We have discussed various types of laser pulse shaping methods for coherent control of multi-level atomic systems. Spectral amplitude blocking in the ladder-type three-level two-photon transition in rubidium, and also in the V-type three-level two-photon transition in rubidium, has shown transition enhancements albeit the loss of pulse energy. Even in a simple two-level two-photon transition, coherent controls with spectral chirps have enhanced the transition in the strong field interaction case, due to the dynamic Stark shift, contradicting the common wisdom of nonlinear optics. Furthermore, the use of higher order chirps for the two-photon transition in a V-type three-level system has mimicked coherent transient-like behaviors. Ultimately, it has been found that optimal pulse shapes can be systematically designed by proper phase function programming for those transitions whose transition phase information is a priori known for all transition pathways. It is hoped that the pulse shaping method finds its versatile usage of ultra-short and ultra-broadband laser field not only for monitoring short time dynamics but also for surgically selective and diverse ways of quantum system controls.

References

1. C.J. Bardeen et al., Feedback quantum control of molecular electronic population transfer. Chem. Phys. Lett. **280**, 151–158 (1997)
2. A. Assion et al., Control of chemical reactions by feedback-optimized phase-shaped femtosecond laser pulses. Science **282**, 919–922 (1998)
3. D. Meshulach, Y. Silberberg, Coherent quantum control of two-photon transitions by a femtosecond laser puse. Nature **396**, 239–242 (1998)
4. G. Imeshev, M.A. Arbore, M.M. Fejer, A. Galvanauskas, M. Fermann, D. Harter, Ultrashort-pulse second-harmonic generation with longitudinally nonuniform quasi-phase-matching gratings: pulse compression and shaping. J. Opt. Soc. Am. B. **17**, 304–318 (2000)
5. S. Cialdi, M. Petrarca, C. Vicario, High-power third-harmonic flat pulse laser generation. Opt. Lett. **31**, 2885–2887 (2006)
6. N. Dudovich, B. Dayan, S.M. Gallagher-Faeder, Y. Silberberg, Transform-limited pulses are not optimal for resonant multiphoton transitions. Phys. Rev. Lett. **86**, 47–50 (2001)
7. S. Lee, J. Lim, C.Y. Park, J. Ahn, Strong-field quantum control of 2 + 1 photon absorption of atomic sodium. Opt. Express **19**, 2266–2277 (2011)
8. B. Chang. Chen, S.H. Lim, Optimal laser pulse shaping for interferometric multiplex coherent anti-Stokes Raman scattering microscopy. J. Phys. Chem. B **112**, 3653–3661 (2008)
9. J.P. Ogilvie et al., Use of coherent control for selective two-photon fluorescence microscopy in live organisms. Opt. Express **14**, 759–766 (2006)
10. C. Rangan, P.H. Bucksbaum, Optimally shaped terahertz pulses for phase retrieval in a Rydberg-atom data register. Phys. Rev. A **64**, 033417 (2001)
11. B.E. Cole, J.B. Williams, B.T. King, M.S. Sherwin, C.R. Stanley, Coherent manipulation of semiconductor quantum bits with terahertz radiation. Nature **410**, 60–63 (2001)
12. W.S. Warren, H. Rabitz, M. Dahleh, Coherent control of quantum dynamics: the dream is alive. Science **259**, 1581–1589 (1993)
13. S. Lee, J. Lim, J. Ahn, Strong-field two-photon absorption in atomic cesium: an analytical control approach. Opt. Express **17**, 7648–7657 (2009)
14. A.M. Weiner, Femtosecond pulse shaping using spatial light modulators. Rev. Sci. Instrum. **71**, 1929 (2000)
15. A.M. Weiner, Ultrafast optical pulse shaping: a tutorial review. Opt. Comm. **284**, 3669–3692 (2011)
16. P. Tournois, Acousto-optic programmable dispersive filter for adaptive compensation of group delay time dispersion in laser systems. Opt. Comm. **140**, 245–249 (1997)
17. F. Verluise et al., Amplitude and phase control of ultrashort pulses by use of an acousto-optic programmable dispersive filter: pulse compression and shaping. Opt. Lett. **25**, 575–577 (2000)
18. T. Brixner, G. Gerber, Femtosecond polarization pulse shaping. Opt. Lett. **26**, 557–559 (2001)
19. J.C. Diels, W. Rudolph, *Ultrashort Laser Pulse Phenomena*, (Academic press, 1996), pp. 1–4
20. J. Lim, K. Lee, J. Ahn, Ultrafast Rabi flopping in a three-level energy ladder. Opt. Lett. **37**, 3378–3380 (2012)
21. J. Lim, H. Lee, S. Lee, J. Ahn, Quantum control in two-dimensional Fourier-transform spectroscopy. Phys. Rev. A. **84**, 013425 (2011)
22. J. Lim, H. Lee, J. Kim, S. Lee, J. Ahn, Coherent transients mimicked by two-photon coherent control of a three-level system. Phys. Rev. A. **83**, 053429 (2011)
23. A.M. Weiner, J.P. Heritage, E.M. Kirschner, High-resolution femtosecond pulse shaping. J. Opt. Soc. Am. B. **5**, 1563 (1988)
24. L. Xu et al., Programmable chirp compensation for 6-fs pulse generation with a prism-pair-formed pulse shaper. IEEE J. Quantum Electron. **36**, 893 (2000)
25. M.M. Wefers, K.A. Nelson, Programmable phase and amplitude femtosecond pulse shaping. Opt. Lett. **18**, 2032 (1993)
26. J.W. Wilson, P. Schlup, R.A. Bartels, Ultrafast phase and amplitude pulse shaping with a single, one-dimensional, high-resolution phase mask. Opt. Express. **15**, 8979 (2007)

27. A.M. Weiner, Femtosecond optical pulse shaping and processing. Prg. Quant. Electr. **19**, 161–237 (1995)
28. M.M. Wefers, K.A. Nelson, Analysis of programmable ultrashort waveform generation using liquid-crystal spatial light modulators. J. Opt. Soc. Am. B. **12**, 1343 (1995)
29. C. Tralleor-Herrero et al., Coherent control of strong field multiphoton absorption in the presence of dynamic Stark shifts. Phys. Rev. A. **71**, 013423 (2005)
30. B. Chatel, J. Degert, S. Stock, B. Girad, Competition between sequential and direct paths in a two-photon transition. Phys. Rev. A. **68**, 041402(R) (2003)
31. C. Trallero-Herrero, J.L. Cohen, T. Weinacht, Strong-field atomic phase matching. Phys. Rev. Lett. **96**, 063603 (2006)
32. S. Lee, J. Lim, J. Ahn, Strong-field two-photon transition by phase shaing. Phys. Rev. A. **82**, 023408 (2010)
33. S. Zamith, J. Degert, S. Stock, B.D. Beauvoir, Observation of coherent transients in ultrashort chirped excitation of an undamped two-level system. Phys. Rev. Lett. **87**, 033001 (2001)
34. J. Degert, W. Wohlleben, B. Chatel, M. Motzkus, B. Girad, Realization of a time-domain Fresnel lens with coherent control. Phys. Rev. Lett. **89**, 203003 (2002)
35. N. Dudovich, D. Oron, Y. Silberberg, Coherent transient enhancement of optically induced resonant transitions. Phys. Rev. Lett. **88**, 123004 (2002)

Chapter 5
Photo-Electron Momentum Spectra in Strong Laser-Matter Interactions

Armin Scrinzi

Abstract The momentum distribution of photo-electrons emitted in the interaction of strong laser pulses with matter contains detailed information about electronic and nuclear structure, electronic correlation, and dynamics of the laser-matter interaction. Beyond the simplest models, correct interpretation of the spectra requires large scale simulations. In particular, at longer, near infra-red wavelength, most calculations are restricted to models with a single active electron. In this chapter the fundamental reasons for this difficulty will be analyzed. The t-SURFF method that allows to overcome those limitations will be explained and illustrated with examples of fully differential single- and double-electron emission spectra at infrared and extreme ultraviolet wavelength.

5.1 Introduction

Differential double photo-electron spectra and the corresponding ionic recoil momentum spectra testify of dynamical correlation between the electrons. By sweeping extreme ultraviolet (XUV) photon energies from below to above the threshold for single-photon double ionization of the He atom one probes correlation in initial and final states. At wavelength in the infrared (IR) range, momentum distributions of recoil ions provide evidence for the importance of re-collision processes [1], where first one electron is ionized, which subsequently is re-directed by the oscillating laser field into a collision with its parent ion causing excitation and possibly detachment of the second electron. The early observation of unexpectedly enhanced double ionization of helium in IR fields [2] is now generally ascribed to this mechanism. Experimental data on strong field IR photo-ionization is also available for many other

---

A. Scrinzi (✉)
Ludwig Maximilians University, Munich, Germany
e-mail: armin.scrinzi@lmu.de

K. Yamanouchi et al. (eds.), *Progress in Ultrafast Intense Laser Science XI*,
Springer Series in Chemical Physics 109, DOI: 10.1007/978-3-319-06731-5_5,
© Springer International Publishing Switzerland 2015

atomic and molecular systems and it was even proposed to use re-collision electron spectra for the analysis of the structure and dynamics of molecules [3].

In a broad range of recent experiments, strong IR laser pulses, often combined with high harmonic pulses, are used to study the electronic dynamics of atoms and molecules on the natural time scale of valence electron motion $\lesssim 1$ fs. Basic mechanisms of the IR-electron interaction are well understood within the simple semi-classical re-collision model [1], but for a more detailed understanding numerical simulations must be employed. This is due to the fundamentally non-perturbative interaction of near IR fields with the valence electrons at intensities of $\gtrsim 10^{14}$ W/cm^2. Even for the simplest single-electron models the simulation remains challenging, especially when accurate photo-electron momentum spectra are required, as, e.g., for re-collision imaging [3–5]. When two-electron processes are involved, one quickly reaches the limits of present day computer resources [6, 7].

For the XUV wave-length, theoretical and experimental questions have matured, even if still under debate (see, e.g., [8] for a recent contribution to the debate with ample references to theory and experiment). At longer wavelength, the large body of experimental data (see, e.g., [9–13] and references therein) and the somewhat smaller range of theoretical models largely based on classical or semi-classical methods (see, e.g., [14–17]) are all plagued by the almost complete lack of reliable theoretical verification, with the notable exception of a few very large scale simulations of two-electron systems in strong fields [18–20], where, however, only in [20] the full two-electron dynamics is treated for laser wavelength of 800 nm.

There are several reasons for this striking absence of complete ab initio simulations of ionization of two-electron systems. Firstly, even the effort for computing single ionization grows dramatically with λ due to the "infrared curse" discussed below. When the effect of the field is perturbative, the situation for single ionization relaxes somewhat, as one basically only needs to know the initial neutral state and the single-electron stationary scattering solutions in the energy range of interest. Although obtaining scattering solutions may be difficult, there is a well defined procedure and the whole technology of electron-ion scattering theory available to approach the problem. For double ionization, also in the perturbative regime the situation is more complex. The convenient partition into bound and singly ionized spectral eigenfunctions cannot be continued to above the double-ionization threshold: eigenfunctions above the double-ionization threshold will in general have, both, single- and double-ionization asymptotics. For distinguishing single from double ionization we therefore invariably need the solutions at large distances. Even without the need for asymptotic analysis, scattering with open double ionization channels is a challenging task for theory.

Numerical simulations, in principle, can provide the full answer. The asymptotic analysis is usually done by propagating the wave function until after the end of the pulse and then analyzing its remote parts either in terms of momentum eigenfunctions, i.e. plane waves, or, when the tails of the ionic Coulomb potentials are not considered negligible, in terms of two-body Coulomb scattering solutions. Three-body scattering solutions, which would obviate a purely asymptotic analysis,

5 Photo-Electron Momentum Spectra in Strong Laser-Matter Interactions

unfortunately, are not accessible. The by far largest part of the computational effort in these simulations goes into following the solution to large distances until the pulse is over and analysis can begin.

5.1.1 Size of the Computational Problem: The Infrared Curse

There are two independent reasons for the surprising difficulty in simulating a seemingly simple process like ionization of an atom or molecule by a dipole laser field: the "curse of dimensions" and the "infrared curse". The curse of dimensions refers to the exponential growth of problem size with the number of degrees of freedom involved in a process. For field free *bound* states of Helium (6 degrees of freedom), problem size can be reduced to the diagonalization of a $\mathcal{O}(100) \times \mathcal{O}(100)$ matrix by exploiting symmetries and using smart coordinate systems. For field-free two-electron *scattering* states the problem remains a large one and is still an area of active numerical development even in the perturbative regime, where only field-free states are required.

These problems are aggravated by the *infrared curse*, i.e. the extremely unfavorable scaling of problem size with the laser wavelength λ. The scaling is λ^2 for each spatial direction of each electron and $\sim\lambda$ for increase in pulse duration, leading to a horrendous scaling as $\sim\lambda^{13}$ for the complete problem. When restricting to linear polarization, the increase is mostly in polarization direction, i.e. 2 out of the 6 coordinates, but still the scaling would be $\sim\lambda^5$.

The scaling in time is simply dictated by the increase of the optical cycle, while, of course, the fundamental time-scale of atomic motion remains fixed: the single cycle at Ti:Sapphire wavelength of $\lambda = 800\,\text{nm}$ lasts $\gtrsim 2.5\,\text{fs}$ or about 110 atomic units ($\text{a.u.}, \hbar = e^2 = m_e = 1$).

The scaling in space is due to a simultaneous increase in wavefunction size and maximum electron momenta. The distance x_{max} traveled by a photo-electron during the pulse grows with pulse duration $\propto \lambda$. This sets a lower limit for the required simulation box-size, if reflections from box boundaries are to be avoided. For example, at an energy of $13\,\text{eV} \approx 1/2\,\text{a.u.}$ the electron moves to ≈ 110 Bohr during a single optical cycle. Usually higher energies and longer pulse durations including the rise and fall of the pulses are of interest, leading to simulation volumes with diameters of thousands of atomic units. The energies reached by an electron in the field are at least $2\,U_p$ for the "direct" and extend further up to about $10\,U_p$ for the "re-scattered" electrons. Re-scattered electrons are those that after ionization are re-directed to the nucleus by the laser field, where they absorb more photons in an inelastic scattering process. The ponderomotive potential $U_p = I/(4\omega^2)$ at laser frequency ω grows linearly with laser intensity I and quadratically with wavelength. For representing corresponding momenta $\lesssim p_{\text{max}} \propto \lambda\sqrt{I}$ we need grid spacings of at least $\Delta x \lesssim 2\pi/p_{\text{max}}\,\text{a.u.}$. The total number of discretization points is then $x_{\text{max}}/\Delta x \propto \lambda^2$. This leaves us with thousands of grid points in each spatial direction even for moderate laser parameters such as $I = 10^{14} \, \text{W/cm}^2$ and $\lambda = 800 \, \text{nm}$ with $U_p = 0.22 \, \text{a.u.} \approx 6 \, \text{eV}$.

The general requirement on discretization cannot be overcome by any specific representation of the wave function: speaking in terms of classical mechanics, we must represent the phase space that is covered by the electrons, which involves a certain range of momenta and positions. If we have no additional knowledge of the structure of solution, the number of discretization points we need is the phase space volume divided by the Planck constant h. In some cases like, for example, single-photon ionization, we can exploit the fact that at long distances the solution covers only a very narrow range of momenta and only the spatially well-localized initial bound state requires a broader range of momenta: the phase-space volume remains small, simple models like perturbation theory allow reproducing the physics. We have no such simplifying physical insight for strong-field IR photo-ionization.

The lower limit for the number of discretization points for the complete wave function can be approached by different strategies: the choice of velocity gauge [21], working in the Kramers-Henneberger frame [22] or in momentum space [23], by variable grid spacings, or by expanding into time-dependent basis functions [24]. A promising strategy is to follow the solution in time [25].

The only way to further cut down on the problem size is to discard part of the solution that is known exactly or in good approximation, i.e. discard parts of the wave function far from the nucleus (and at large inter-electron distance) and extract the desired spectral information from the truncated solution. This approach is taken by the time-dependent surface flux method (t-SURFF) introduced in this chapter.

To explain the main idea of t-SURFF we first discuss the single-electron system. Further mathematical detail for its extension to the two-electron case is given next. In its simplest version using Volkov solutions , the method is exact only for finite-range potentials where a full numerical solution is required over whole range of the potential. Errors due to the Coulomb potential can be systematically controlled. With further numerical effort, the errors can be completely eliminated by numerically solving an auxiliary single-particle scattering problem.

We discuss accuracies and demonstrate the efficiency of t-SURFF for calculating photo-electron momentum spectra for single-electron systems in three dimensions. Shake-up and double ionization of a two-electron system are calculated for a standard two-dimensional soft-core model as well as Helium in full three dimensional space.

5.2 The t-SURFF Method

5.2.1 Single-Electron Systems

Scattering measurements and theory are both based on the fact that interactions are limited to finite ranges in space and time and that at large distances R_c and large times T the time-evolution of the scattering particle is that of free motion:

5 Photo-Electron Momentum Spectra in Strong Laser-Matter Interactions

$$\Psi(\mathbf{r}, t) \sim \int dk^{(3)} \exp(-it\mathbf{k}^2/2)b(\mathbf{k})\chi_{\mathbf{k}}(\mathbf{r}) \qquad \text{for } t > T, |\mathbf{r}| > R_c, \qquad (5.1)$$

where $\chi_{\mathbf{k}}(\mathbf{r}) = (2\pi)^{-3/2}\exp(i\mathbf{k} \cdot \mathbf{r})$ are δ-normalized plane waves. The measured momentum spectral density is the square of the spectral amplitude $b(\mathbf{k})$

$$\sigma(\mathbf{k}) = |b(\mathbf{k})|^2. \qquad (5.2)$$

The spectral amplitudes $b(\mathbf{k})$, in turn, are obtained from the solution at T as

$$b(\mathbf{k}) = \int\limits_{|\mathbf{r}| > R_c} d^{(3)}r \, e^{it\mathbf{k}^2/2}\chi_{\mathbf{k}}^*(\mathbf{r})\Psi(\mathbf{r}, T), \qquad (5.3)$$

Note that T, in general, must be chosen much larger than the pulse duration such that the slower electrons can reach the asymptotic region.

This fundamental principle of scattering theory is the main prerequisite for using t-SURFF: there is some radius R_c beyond which the Hamiltonian exactly or approximately reduces to an asymptotic one with known solutions. Let $H(t)$ denote the time dependent Hamiltonian of our system and assume that there exists an exactly solvable $H_v(t)$ that at large distances agrees with $H(t)$:

$$H_v(t) = H(t) \quad \text{for } |\mathbf{r}| > R_c \text{ and } \forall t. \qquad (5.4)$$

For a single particle in a laser field described in velocity gauge and a short range potential $V(\mathbf{r}) \equiv 0$ for $|\mathbf{r}| > R_c$, $H_v(t)$ is the Hamiltonian for the free motion in the laser field

$$H_v(t) = \frac{1}{2}[-i\nabla - \mathbf{A}(t)]^2, \qquad (5.5)$$

where $\mathbf{A}(t) = -\int_{-\infty}^{t} \mathscr{E}(t')dt'$ for an electric dipole field $\mathscr{E}(t)$. Here and throughout we use atomic units $\hbar = m_e = e^2 = 1$ unless indicated otherwise. Electron mass and unit charge are denoted by m_e and e, respectively, the Bohr radius results as the atomic unit of length. The TDSE with $H_v(t)$ has the Volkov solutions

$$\chi_{\mathbf{k}}(\mathbf{r}, t) = (2\pi)^{-3/2}e^{-i\Phi(\mathbf{k}, t)}e^{i\mathbf{k}\cdot\mathbf{r}}, \qquad (5.6)$$

i.e. plane waves where all time-dependence is in the well-known Volkov phase

$$\Phi(\mathbf{k}, t) = \frac{1}{2}\int\limits_{-\infty}^{t} dt'[\mathbf{k} - \mathbf{A}(t')]^2. \qquad (5.7)$$

100 A. Scrinzi

We now choose T and R_c large enough such that at T the pulse is over

$$H_v(t) = -\frac{1}{2}\Delta \quad \text{for } t > T \tag{5.8}$$

and the wave function has split into its bound and scattering parts

$$\Psi(\mathbf{r}, T) = \Psi_b(\mathbf{r}, T) + \Psi_s(\mathbf{r}, T) \tag{5.9}$$

with the properties

$$\Psi_b(\mathbf{r}, T) \approx 0 \quad \text{for } |\mathbf{r}| \geq R_c \tag{5.10}$$

$$\Psi_s(\mathbf{r}, T) \approx 0 \quad \text{for } |\mathbf{r}| \leq R_c. \tag{5.11}$$

The approximate sign in (5.10) refers to the tails of any bound state function, which decay exponentially with increasing R_c. The approximate sign in (5.11) refers to the fact that electrons with very low energies $\mathbf{k}^2/2 \sim 0$ may not have passed R_c at time T. Only $\Psi_s(\mathbf{r}, T)$ contributes to the photo-electron spectrum. As it vanishes inside the radius R_c, we can multiply the full $\Psi(\mathbf{r}, T)$ by the function

$$\theta(\mathbf{r}, R_c) = \begin{cases} 0 & \text{for } |\mathbf{r}| < R_c \\ 1 & \text{for } |\mathbf{r}| \geq R_c \end{cases} \tag{5.12}$$

write the emission amplitude for photo-electron momentum $\mathbf{k}$ as

$$b(\mathbf{k}, T) = \int d^{(3)}r\, \chi_k^*(\mathbf{r}, T)\theta(\mathbf{r}, R_c)\Psi(\mathbf{r}, T) =: \langle \chi_{\mathbf{k}}(T)|\theta(R_c)|\Psi(T)\rangle. \tag{5.13}$$

To convert the volume integral into a time integral over the volume's surface, we write (5.13) as an integral of the time derivative and use the fact that on the support of $\theta(\mathbf{r}, R_c)$ both, $\chi_{\mathbf{k}}(\mathbf{r}, t)$ and $\Psi(\mathbf{r}, t)$ evolve by the same Hamiltonian $H_v(t)$. We obtain

$$\langle \chi_k(T)|\theta(R_c)|\Psi(T)\rangle = i\int_0^T dt\langle \chi_k(t)|[H_v(t), \theta(R_c)]|\Psi(t)\rangle, \tag{5.14}$$

where the commutator

$$[H_v(t), \theta(R_c)] = \left[-\frac{1}{2}\Delta + i\mathbf{A}(t)\cdot\nabla, \theta(R_c)\right] \tag{5.15}$$

vanishes everywhere except on $|\mathbf{r}| = R_c$: the asymptotic information is obtained by integrating the time dependent flux through a surface at finite distance R_c (see Fig. 5.1).

5 Photo-Electron Momentum Spectra in Strong Laser-Matter Interactions

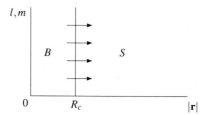

Fig. 5.1 t-SURFF for a single electron: electrons pass R_c from the "bound" region B to the "singly ionized region" S for all angular degrees of freedom l, m. Ionization spectra are obtained by integrating the flux through the boundary. For linear polarization the m-quantum number is conserved

The idea of replacing one dimension of the volume integral by a time integration had been used for reactive scattering with time-independent Hamiltonians. However, photo-ionization cannot be computed using those methods, as the dipole interaction is non-local and the external field modifies the particle energies everywhere, in particular also after the particle has passed the radius R_c. Without time-dependence we can make one more step, as then the Volkov phase reduces to $\Phi(\mathbf{k}, t) = t\mathbf{k}^2/2$ and time-integration turns into the time-energy Fourier-transform of the surface integral, which connects the present approach to [26, 27].

5.2.2 Single-Ionization into Multiple Ionic Channels

Next we discuss ionization of few electron systems. The simplest extension of the single electron method is for computing single ionization into ground and excited state ionic channels. The complication is that in presence of a strong field the ionic state can differ from the field free ionic state due to polarization and one must make sure to count flux passing the surface into the correct ionic channel. In this section we consider only the lowest ionic states that remain bound and do not contribute to double ionization. For simplicity, all formulae are written for a two-electron system. However, replacing the ionic states with few-electron ionic states, immediately generalizes the approach to the few-electron case.

For decomposing coordinate space into bound and asymptotic regions we define projector functions $\theta_1(\mathbf{r}_1, R_c)$ and $\theta_2(\mathbf{r}_2, R_c)$ for both coordinates $\mathbf{r}_1$ and $\mathbf{r}_2$, respectively, analogous to the single particle case (5.12). Again picking sufficiently large times T and a sufficiently large surface radius R_c we can partition the wave function $\Psi(T)$ into its bound (Ψ_b), singly ionized ($\Psi_s, \Psi_{\bar{s}}$), and doubly ionized (Ψ_d) parts (see Fig. 5.2)

$$\Psi(T) = (1 - \theta_1)(1 - \theta_2)\Psi(T)$$
$$+ \theta_1(1 - \theta_2)\Psi(T) + (1 - \theta_1)\theta_2\Psi(T) + \theta_1\theta_2\Psi(T) \quad (5.16)$$
$$:= \Psi_b(T) + \Psi_s(T) + \Psi_{\bar{s}}(T) + \Psi_d(T) \quad (5.17)$$

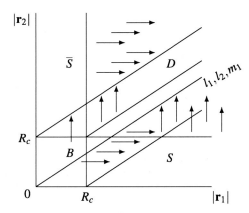

Fig. 5.2 Partitioning of coordinate space into bound (B), singly ionized (S, $\bar{S}$) and doubly ionized D regions. Single and double photo electron spectra are obtained by integrating the flux across the boundaries between the regions. The *axis* into the drawing plane symbolizes the angular degrees of freedom $l_1, l_2, m_1 = M - m_2$, assuming linear polarization, i.e. conserved total M-quantum number

Here and in the following we suppress the arguments $\mathbf{r}_1$, $\mathbf{r}_2$ and R_c of θ_1 and θ_2. Note that the assignment of singly and doubly ionized character to the different regions is asymptotically exact: the error can be made arbitrarily small for any specific solution $\Psi(T)$ by choosing sufficiently large T and R_c.

The Hamiltonian for a single-ionization channel

$$H_s(t) = H_v(t) \otimes H_{ion}(t) \tag{5.18}$$

agrees with the exact Hamiltonian on the support of $\theta_1(1 - \theta_2)$ (region S in Fig. 5.2). The corresponding Hamiltonian on $\bar{S}$ is obtained by particle exchange. Channel solutions $\chi_{c,\mathbf{k}}(\mathbf{r}_1, \mathbf{r}_2, t)$ for the TDSE on S

$$i\frac{d}{dt}\chi_c(\mathbf{r}_1, \mathbf{r}_2, t) = H_s(t)\chi_c(\mathbf{r}_1, \mathbf{r}_2, t) \tag{5.19}$$

have the form

$$\chi_{c,\mathbf{k}}(\mathbf{r}_1, \mathbf{r}_2, t) = \chi_\mathbf{k}(\mathbf{r}_1) \otimes \phi_c(\mathbf{r}_2, t), \tag{5.20}$$

where $\chi_\mathbf{k}(\mathbf{r}_1)$ is given by (5.6). The $\chi_{c,\mathbf{k}}$ describe the asymptotic solutions of the *non-interacting* system into which we choose to decompose our *interacting* solution Ψ_s in the area S. Here $\phi_c(\mathbf{r}_2, t)$ solves the ionic TDSE

$$i\frac{d}{dt}\phi_c(\mathbf{r}_2, t) = H_{ion}(t)\phi_c(\mathbf{r}_2, t). \tag{5.21}$$

5 Photo-Electron Momentum Spectra in Strong Laser-Matter Interactions 103

At large times $t > T$, the ionic factor $\phi_c(\mathbf{r}_2, T)$ should correspond to an ionic state of our choice. Therefore, rather than an initial condition, we use a *final* condition for the time evolution of ϕ_c (we need to solve the ionic TDSE backward in time). The channel solution on the support $\overline{S}$ of the particle exchanged projector $(1 - \theta_1)\theta_2$ is $\overline{\chi}_{c,\mathbf{k}}(\mathbf{r}_1, \mathbf{r}_2, t) = \chi_{c,\mathbf{k}}(\mathbf{r}_2, \mathbf{r}_1, t)$.

With the spectral amplitude

$$b(\mathbf{k}, c, T) = \langle \chi_{\mathbf{k},c}(T)|\theta_1(1 - \theta_2)|\Psi_s(T)\rangle. \tag{5.22}$$

the probability density for finding at time T an electron with momentum $\mathbf{k}$ in ionic channel c is

$$\sigma(\mathbf{k}, c, T) = 2|b(\mathbf{k}, c, T)|^2. \tag{5.23}$$

The factor 2 arises from adding the two identical exchange symmetric contributions. For converting this integral into a time integral over a surface we make the simplifying assumption that the ionic solution never leaves the bound area

$$\phi_c(\mathbf{r}_2, t) \approx 0 \quad \text{for } |\mathbf{r}| > R_c \quad \text{and} \quad \forall t. \tag{5.24}$$

This is the precise meaning of the assumption that the ionic states considered does not get further ionized. The approximate sign refers to the fact that again there is always an exponential tail reaching to arbitrary distances and further that interaction with any pulse will lead to a small amount of ionization. By the condition (5.24) we neglect the flux between the singly ionized regions S, $\overline{S}$ and the doubly ionized region D at all times. Then, using the same procedure as for a single particle we obtain

$$b(\mathbf{k}, c, T) = i \int_{-\infty}^{T} dt \langle \chi_{\mathbf{k}}(t)|[H_v(t), \theta_1]|\psi_c(t)\rangle \tag{5.25}$$

with the auxiliary channel-projected wave function $\psi_c(\mathbf{r}_1, t)$ defined as

$$\psi_c(\mathbf{r}_1, t) := \int d^3r_2 \phi_c^*(\mathbf{r}_2, t)\Psi(\mathbf{r}_1, \mathbf{r}_2, t). \tag{5.26}$$

For evaluating the integral (5.25) we only need to know values and radial derivatives of ψ_c on the surface $|\mathbf{r}| = R_c$.

For the computation it means solving the full two-electron problem up to time T and up to radius R_c. Beyond R_c one can absorb all amplitudes. In addition, for each channel c, we need to solve one single electron problem up to the same time and radius (which is usually a much simpler task).

5.2.3 Double Ionization Spectra

When double ionization occurs, flux passes from the bound region B through the singly ionized regions S or $\overline{S}$ into the doubly ionized region D. Our naming of the areas is suggestive but does not imply any bias as to an actual state that the electrons occupy within any of these areas. It is unsubstantial for the present discussion whether some intermediate ionic bound state is occupied by one of the electrons in S or $\overline{S}$ (sequential ionization) or whether both electrons must be considered unbound. The sole purpose of the partitioning is to have well-defined surfaces outside the ranges of the respective potentials where we will integrate fluxes.

For double ionization there is one obvious limitation of the discussion so far: on the line $|\mathbf{r}_1 - \mathbf{r}_2| = a$ the electron-electron interaction is constant and not negligible for small a. This is a general problem for any multi-particle breakup, which makes break-up processes more complex than single particle scattering. A pragmatic solution has been effectively employed in many earlier publications, which is to neglect electron-electron repulsion at large distances from the nucleus. Any projection onto products of single electron states effectively makes this approximation, be it products of Coulomb scattering waves or of plane waves (both approaches are discussed, for example, in [28]). Sensitivity to this approximation can be tested by varying the distance from the nucleus where the projection starts. As that solution was found to work well in many cases, we make this approximation explicit in the present chapter by smoothly turning off all potentials including the electron-electron interaction before the surface radius R_c. In that case we can always use the free (Volkov) Hamiltonian $H_v(t)$ beyond R_c.

By our assumptions, in the region $S : |\mathbf{r}_1| \geq R_c$, the Hamiltonian is identical to $H_s(t)$, (5.18), which motivates the ansatz

$$\theta_1 \Psi(\mathbf{r}_1, \mathbf{r}_2, t) = \int d^3k \sum_n \chi_\mathbf{k}(\mathbf{r}_1, t)\xi_n(\mathbf{r}_2)a(\mathbf{k}, n), \tag{5.27}$$

with the Volkov solutions $\chi_\mathbf{k}(\mathbf{r}_1, t)$ on coordinate $\mathbf{r}_1$ and an expansion into a time-independent, complete, but otherwise arbitrary set of functions $\xi_n(\mathbf{r}_2)$ on $\mathbf{r}_2$. Using orthogonal projection onto the expansion functions $\chi_{\mathbf{k}_1}$ and ξ_n, the coefficients $a(\mathbf{k}_1, n, t)$ are obtained as

$$a(\mathbf{k}_1, n, t) = \int d^3k_1' q_\theta(\mathbf{k}_1, \mathbf{k}_1', t)b(\mathbf{k}_1', n, t) \tag{5.28}$$

with

$$b(\mathbf{k}, n, t) = \int_{-\infty}^{\infty} d^3r_1 \chi_{\mathbf{k}_1'}^*(\mathbf{r}_1, t)\theta_1 \int_{-\infty}^{\infty} \xi_n^*(\mathbf{r}_2)\Psi(\mathbf{r}_1, \mathbf{r}_2, t). \tag{5.29}$$

5 Photo-Electron Momentum Spectra in Strong Laser-Matter Interactions 105

The integral (5.28) over q_θ accounts for the fact that the plane waves are not δ-orthonormal when the integration is restricted by θ_1: the inverse overlap q_θ is defined by

$$\int d^3k' q_\theta(\mathbf{k}, \mathbf{k}', t)\langle \mathbf{k}', t|\theta_1|\mathbf{k}'', t\rangle = \delta^{(3)}(\mathbf{k} - \mathbf{k}'').$$ (5.30)

For notational brevity, we assume here that the basis functions ξ_n are orthonormal. The time-derivative of the $b(\mathbf{k}_1, n, t)$ is

$$i\frac{d}{dt}b(\mathbf{k}_1, n, t) = \langle \chi_{\mathbf{k}_1}(t)|\theta_1\langle \xi_n|H_2(t)|\Psi(t)\rangle\rangle$$
$$- \langle \chi_{\mathbf{k}_1}(t)|[H_v(t), \theta_1]\langle \xi_n|\Psi(t)\rangle\rangle.$$ (5.31)

By the double right bracket $\rangle\rangle$ we emphasize that integration is over both coordinates, $\mathbf{r}_1$ and $\mathbf{r}_2$. Inserting the representation (5.27) into the first term we obtain an inhomogeneous equation for the $b(\mathbf{k}_1, n, t)$:

$$i\frac{d}{dt}b(\mathbf{k}_1, n, t) = \sum_m \langle \xi_n|H_{ion}(t)|\xi_m\rangle b(\mathbf{k}_1, m, t)$$
$$- \langle \mathbf{k}_1, t|[H_v(t), \theta_1]\langle \xi_n|\Psi(t)\rangle\rangle$$ (5.32)

where we have used (5.30). The inhomogeneity is the flux through the surface $|\mathbf{r}_1| = R_c$. Initial conditions are $b(\mathbf{k}_1, n) \equiv 0$, i.e. no electrons outside R_c. We write the double ionization amplitude

$$b(\mathbf{k}_1, \mathbf{k}_2, T) = \langle \chi_{\mathbf{k}_2}(T)|\langle \chi_{\mathbf{k}_1}(T)|\theta_2\theta_1|\Psi(T)\rangle\rangle,$$ (5.33)

and use one more time the conversion to integrals over surface flux

$$b(\mathbf{k}_1, \mathbf{k}_2, T) = \langle \chi_{\mathbf{k}_2}(T)|\langle \chi_{\mathbf{k}_1}(T)|\theta_2\theta_1|\Psi(T)\rangle\rangle$$ (5.34)

$$= \int_0^T dt \frac{d}{dt}\langle \chi_{\mathbf{k}_2}(t)|\langle \chi_{\mathbf{k}_1}(t)|\theta_2\theta_1|\Psi(t)\rangle\rangle$$ (5.35)

$$=: i\int_0^T dt[B(\mathbf{k}_1, \mathbf{k}_2, t) + \overline{B}(\mathbf{k}_1, \mathbf{k}_2, t)].$$ (5.36)

The two terms B and $\overline{B}$ are related by exchange symmetry

$$B(\mathbf{k}_1, \mathbf{k}_2, t) = \overline{B}(\mathbf{k}_2, \mathbf{k}_1, t) = \langle \chi_{\mathbf{k}_2}(t)|\langle \chi_{\mathbf{k}_1}(t)[H_v(t), \theta_2]\theta_1|\Psi(t)\rangle\rangle$$ (5.37)

For computing B, we only need to know $\theta_1 \Psi(t)$, for which we insert the representation (5.27) to obtain

$$B(\mathbf{k}_1, \mathbf{k}_2, t) = \sum_m \langle \chi_{\mathbf{k}_2}(t) | [H_v(t), \theta_2] | \xi_m \rangle b(\mathbf{k}_2, m, t) \qquad (5.38)$$

The inverse overlap $q_\theta(\mathbf{k}_1, \mathbf{k}_1')$, (5.30), cancels with the overlap integrals and never needs to be evaluated explicitly.

$B(\mathbf{k}_1, \mathbf{k}_2, t)$ is the contribution to the double ionization spectrum passing at time t through surface $|\mathbf{r}_2| = R_c$ from region S into D. In S, the first electron is already detached and has a fixed canonical momentum $\mathbf{k}_1$ that is carried into the double-ionized region D. $\overline{B}(\mathbf{k}_1, \mathbf{k}_2, t)$ is the alternate contribution going through region $\overline{S}$.

5.2.3.1 General Single Ionization Spectra

Once we have obtained the $b(\mathbf{k}_1, n, T)$ we can reconstruct the wave function for $|\mathbf{r}_1| > R_c$. In particular, the amplitude for single ionization spectra for any ionic state ϕ_c is

$$b(\mathbf{k}, c) = \sum_n b(\mathbf{k}, n) a_n^{(c)}, \qquad (5.39)$$

where $a_n^{(c)}$ are the expansion coefficients of ϕ_c with respect to the basis functions ξ_n:

$$\phi_c(\mathbf{r}) = \sum_n \xi_n(\mathbf{r}) a_n^{(c)}. \qquad (5.40)$$

This approach is not limited to non-ionizing ϕ_c, as it analyzes ionic population at time T after the end of the pulse. It also offers an alternative approach to the single-ionization for non-ionizing ϕ_c. Note that here we need to solve the inhomogeneous ionic problem (5.32) for each photo-electron momentum $\mathbf{k}$. The total spectrum is a linear combination of the individual contributions from each n. Where it is applicable, the advantage of the single-ionization procedure of Sect. 5.2.2 is twofold: firstly, for each final ϕ_c one needs to solve only one ionic TDSE and compute values and derivatives of the channel surface function ψ_c. The complete spectrum can be obtained by time-integration with little numerical effort. Secondly, as one directly obtains the channel spectrum without intermediated decomposition and final resummation of the wave function, results are in general more robust numerically.

5.2.4 Computational Remarks

The substantial gain of the method is that, rather than computing the full solution in region D and then analyzing it, we only need to integrate the flux through the surface separating S from D. In the direction parallel to that surface the wave function is represented in terms of the free solutions $\chi_{\mathbf{k}_1}(\mathbf{r}_1, t)$, where we do not need to expand the wave function completely, but can restrict propagation to the momenta $\mathbf{k}_1$ that we are interested in. For each $\mathbf{k}_1$ we need to solve equation (5.32), which is an ionic TDSE with an additional source term (the flux entering S from the bound region B).

The approach can only be successful, when absorption does not significantly distort the solution at the integration surfaces. The "infinite range exterior complex scaling" (irECS) absorber was shown in [29, 30] to provide traceless absorption over a very wide energy range at low computational cost. It outperforms standard complex absorbing potentials by several orders of magnitude in accuracy. If needed, irECS can be pushed to full machine precision using not more than 20 discretization coefficients per coordinate in the absorbing region $|\mathbf{r}| > A_0$. Absorption can begin at any $A_0 \geq R_c$.

As mathematical methods, both, t-SURFF and also irECS are independent of any particular numerical discretization. For the computations reported below, we have used a high order finite element discretization, which we find to be computationally convenient. For the time-propagation we have used both, 4th order Magnus exponentiation [31] and also standard 4th order Runge-Kutta. Magnus exponentiation performs better near the perturbative regime, i.e. at low intensities or short wavelength. In the strongly non-perturbative regime, we got best results with classical Runge-Kutta. Note that implicit methods are of no particular use in our situation, as we cannot take advantage of the larger step sizes provided by those methods: we need to sample the wave function at small time steps, if we want to resolve the fast momenta passing through the surface. In most cases, the required step size for the 4th order Runge-Kutta was only about a factor 2 below the time-resolution dictated by high momenta.

All relevant mathematical and numerical details on irECS and our particular version of the finite element method can be found in [29, 30], which also contain several more numerical examples.

5.3 Applications

5.3.1 Spectra for a Short Range Potential

We solve the single-particle time-dependent Schrödinger equation in velocity gauge

$$i\frac{d}{dt}\Psi_v(\mathbf{r}, t) = \left\{\frac{1}{2}[-i\nabla - \mathbf{A}(t)]^2 + V(\mathbf{r})\right\}\Psi_v(\mathbf{r}, t), \qquad (5.41)$$

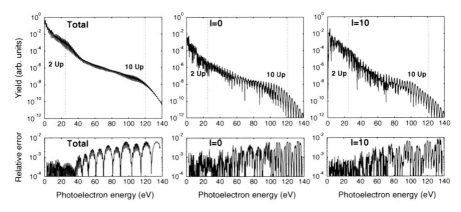

Fig. 5.3 *Upper panels* total and partial wave photo-electron spectra obtained for the smoothly truncated Coulomb potential (5.42). The *lower panels* show the relative errors for each spectrum according to (5.44) of a calculation with only 90 discretization points per angular momentum comparing to a fully converged calculation. *Pulse parameters* $\lambda = 800$ nm, $T = 5$ optical cycles, intensity $= 2 \times 10^{14}$ W/cm^2. (From [30])

where we first use the short-range "Coulomb" potential

$$V(r) = \begin{cases} c\left[-1/r - r^2/(2R^3) + 3/(2R)\right] & \text{for } r \leq R \\ 0 & \text{for } r > R. \end{cases} \quad (5.42)$$

With $R = 20$ and an effective charge $c = 1.1664$ the ground state energy is -0.5. The laser pulse is linearly polarized in z-direction with the vector potential

$$A_z(t) = \frac{\mathscr{E}_0}{\omega} \cos^2\left(\frac{\pi t}{2T}\right) \sin(t\omega). \quad (5.43)$$

We choose parameters $\omega = 0.057$ and $\mathscr{E}_0 = 0.0755$ corresponding to laser wavelength of 800 nm and peak intensity 2×10^{14} W/cm^2. T is the full width at half maximum (FWHM) of the vector potential, total pulse duration is $2T$.

Figure 5.3 shows the total and partial wave photo-electron spectra for potential range $R = 20$ and $T = 5$ optical cycles. At these parameters, more than 90 % of the electrons get detached. For accuracy $\lesssim 1\%$ up to energies of $10\,U_p \approx 120$ eV we need $L_{\max} = 30$ partial waves with only 90 radial discretization points. We define the error relative to an accurate reference spectrum σ_{ref} as

$$\mathscr{D}(E) = \frac{|\sigma(E) - \sigma_{\text{ref}}(E)|}{\max(\sigma(E), \langle \sigma(E) \rangle_{\delta E})}, \quad \langle \sigma(E) \rangle_{\delta E} := \frac{1}{2\delta E} \int_{E-\delta E}^{E+\delta E} dE' \sigma(E'). \quad (5.44)$$

Including in the denominator the average over the interval $[E - \delta E, E + \delta E]$ suppresses spurious spikes in the error due to near-zeros of the spectrum. We choose

5 Photo-Electron Momentum Spectra in Strong Laser-Matter Interactions

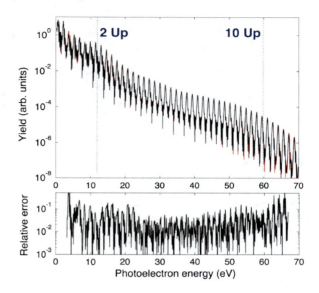

Fig. 5.4 Photo-electron energy spectra for the hydrogen atom (*upper panel*) obtained with surface radius $R_c = 110$ (*black*) and $R_c = 140$ (*red*). *Lower panel* error estimate by comparing the spectra according to (5.44). Pulse parameters $\lambda = 800$ nm, $T = 20$ optical cycles, intensity $= 10^{14}$ W/cm^2. (From [30])

$\delta E = 0.05$ a.u. ≈ 1.5 eV, which at 800 nm corresponds to averaging over about 2 photo-electron peaks.

In the unscaled region we use 60 points, 30 points are located in $|\mathbf{r}| > R_c$. The accuracy estimate shown in Fig. 5.3 is obtained by comparing to a fully converged calculation. When we increase the number of points to 180, the error drops to $\lesssim 10^{-3}$ The increase of relative errors with energy can be attributed to the decrease of the signal: note that from 0 to 10 U_p the spectrum drops by more than 5 orders of magnitude.

5.3.2 Spectra for the Hydrogen Atom

The long-range nature of the Coulomb potential introduces extra mathematical and practical complications. There is no surface radius R_c such that the Volkov solutions become exact. In addition, the Rydberg bound states extend to arbitrarily large distances. Both problems can be controlled with some extra technical effort [30]. Here, we present the simple pragmatic solution of using larger R_c such that the remaining error due to the presence of the Coulomb potential becomes acceptable.

Figure 5.4 shows a spectrum calculated for the Hydrogen atom with a FWHM $T = 20$ optical cycle pulse at 800 nm wavelength and peak intensity 10^{14} W/cm^2. All discretization errors can be controlled in the same way as discussed for the short range potential. The error is dominated by the dependence on the surface radius R_c: on an absolute (logarithmic) scale, two calculations with $R_c = 110$ and $R_c = 140$

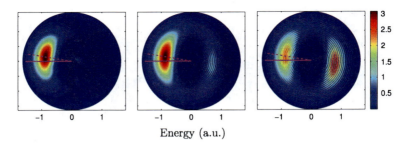

Energy (a.u.)

Fig. 5.5 Photo-emission from single-active-electron Helium at intensity 2×10^{14} W/cm^2, center wavelength 800 nm and near-circular polarization $\mathscr{E}_y/\mathscr{E}_x = \sqrt{5/4}$. Pulse durations are 1 (*left*), 2 (*center*), and 3 (*right*) optical cycles at FWHM. The spectral density is plotted in the polarization (xy-)plane. The distance from the center indicates energy (in atomic units). The *solid magenta line* indicates the direction of the vector potential at *peak* field, the *dashed line* gives the approximate *peak* of emission direction

are hardly discernable. The error level of the calculation with $R_c = 110$ and 180 discretization points is $\lesssim 10\%$ and decreases slowly as R_c increases.

5.3.3 IR Photo-Electron Spectra at Elliptical Polarization

The measurement of photo-emission by elliptically polarized light has renewed a debate about the possibility of measuring tunnel delay times [32]. Apart from its conceptual difficulty, the debate suffers also from the absence of exact solutions to the TDSE. Using t-SURFF we can for the first time present accurate spectra for photo-emission from a single-electron atom at realistic parameters.

Results for a three-dimensional, single-electron model of the Helium atom are shown in Fig. 5.5. Due to the strong non-linearity of emission, for 1 and 2 cycle pulses emission is concentrated near peak field $t_{\text{peak}} : \mathscr{E}(t_{\text{peak}}) = \mathscr{E}_{\text{max}}$ (y-direction). If the electron can be considered free at the instance of peak field, emission direction should peak at $\mathbf{A}_{\text{max}}$, approximately orthogonal to peak $\mathscr{E}_{\text{max}}$. The observed deviation from this direction indicates that the electron continues to interact with the binding potential after t_{peak}. The deflection is largely independent of pulse duration as it dominantly originates from the area close to the nucleus, which the electron leaves briefly after detachment.

5.3.4 Two-Electron System: 2×1-Dimensional Helium

Here we review results from [33], where the extension of t-SURFF to double ionization was demonstrated. We use simple 2×1-dimensional model Hamiltonian

5 Photo-Electron Momentum Spectra in Strong Laser-Matter Interactions

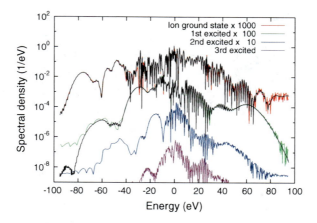

Fig. 5.6 Single photo-electron emission spectra in the lowest 4 ionic channels. Negative energies indicate emission to $(-\infty, -R_c]$. Spectra are calculated using (5.39). In addition, for ground and first excited ion channel, *curves* computed by (5.25) are shown (*black lines*). The *curves* are scaled for better visibility. (From [33])

$$H(t) = \sum_{\alpha=1,2}\left\{\frac{1}{2}\left[-i\frac{\partial}{\partial x_\alpha} - A(t)\right]^2 - \frac{2M(x_\alpha)}{\sqrt{x_\alpha^2 + 1/2}}\right\} + \frac{M(x_1)M(x_2)}{\sqrt{(x_1 - x_2)^2 + 0.3}}, \quad (5.45)$$

where all potentials are switched off outside a finite domain. We use a "truncation radius" C_p and a truncation function $M(x)$ that is $\equiv 1$ up to $|x_\alpha| = C_p - 5$ and goes differentiably smoothly to 0 at $|x_\alpha| = C_p$. In the examples shown we use $C_p = 20$. The screening factor of $1/2$ in the ionic potential was chosen for esthetic reasons, as then the ionic ground state of our one-dimensional model is $E_0 = -2$, as in true, three-dimensional He^+. The first excited ion state occurs at $E_1 = -0.93$, which substantially lower than the first excited He^+ energy of $-1/2$. With the electron-electron screening of 0.3 the ionization potential of the neutral model atom is 0.88. We consider this a fair approximation of the ionization potential of the three-dimensional He (0.90 a.u.).

In all calculations a pulse with $\cos^2$ envelope defined in terms of the vector potential

$$A(t) = \frac{\mathcal{E}_0}{\omega}\cos^2\left(\frac{\pi t}{2T_{FWHM}}\right)\sin(\omega t + \phi_{CEO}), \quad (5.46)$$

where T_{FWHM} is the full-width-half-maximum of the envelope of the vector potential, ω is the laser central photon energy, and ϕ_{CEO} the carrier-envelope offset phase. The peak field amplitude $\mathcal{E}_0$ is related to the pulse peak intensity I by $\mathcal{E}_0 = \sqrt{2I}$. For a "cosine pulse" $\phi_{CEO} = 0$ the peak of the electric field approximately coincides with the maximum of the pulse envelope with minor deviations due to the derivative of the envelope.

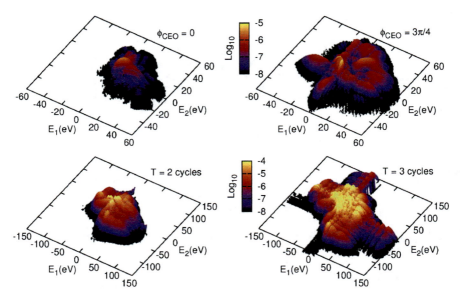

Fig. 5.7 Overview of double ionization spectra for laser wave-length 800 nm and peak intensity 2×10^{14} W/cm^2. *Upper row* pulse duration $T = 1$ optical cycle with $\phi_{CEO} = 0$ (*left*) and $3\pi/4$ (*right*). *Lower row* $\phi_{CEO} = 0$ with $T = 2$ (*left*) and 3 (*right*). Negative energies indicate emission to $(-\infty, -R_c]$. (From [33])

Shake-up by very strong, short IR pulses has been observed recently using "attosecond transient absorption spectroscopy" [34], which can monitor the time-evolution of an excited ionic state after IR ionization.

Within the present simplified model we can only demonstrate basic qualitative features of shake-up in IR ionization. We have studied dependence on the carrier-envelope offset phase ϕ_{CEO} and on pulse duration. We use an intensity of 2×10^{14} W/cm^2 and $\omega = 0.057$ a.u. (corresponding to wavelength 800 nm). Converged results were obtained using the same discretization parameters as in the XUV case discussed above.

In Fig. 5.6 photo electron spectra for the ground and first excited states calculated by formula [(5.25), smooth line] and, alternatively, by (5.39) (coarser energy grid) for a single cycle cosine pulse: $T = 2\pi/\omega$ and $\phi_{CEO} = 0$. The spectra show some generic short pulse features: pronounced asymmetry and absence of individual photo-electron peaks at energies, where emission occurs only during a single laser half-cycle. Emission to the left is lower, as the field amplitude to the left is smaller.

Figure 5.7 gives an overview of doubly-differential photo emission spectra obtained for different ϕ_{CEO} and pulse durations up to three optical cycles $T_{FWHM} = 6\pi/\omega$ a.u. ≈ 7.5 fs. The calculations were performed using a slightly more accurate discretization with 60 points on the half-axis $[0, \infty)$.

As to be expected, we see pronounced asymmetries and strong effects of ϕ_{CEO} for the shortest pulses. While the single-cycle cosine pulse essentially only shows weak

5 Photo-Electron Momentum Spectra in Strong Laser-Matter Interactions 113

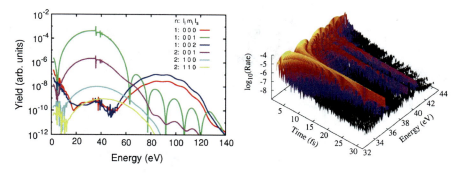

Fig. 5.8 *Left panel* single photo-electron emission spectra in the lowest few ionic shake-up channels. l_i, m_i and l_s denote angular quantum numbers of ion and emitted electron, respectively. *Right panel* build-up rates $d\sigma(E, t)/dt$ of the photo-electron emission spectrum in the $l_i, m_i, l_s = 0, 0, 1$ ionic channel

and unidirectional emission of both electrons, backward emission and significantly higher yields arise at $\phi_{CEO} = 3\pi/4$. Most likely, the responsible mechanism is re-collision.

5.3.5 XUV Photo-Emission from He in Full Dimensionality

Using t-SURFF, at short wavelength, full three-dimensional single- and double-emission from 6-dimensional Helium can be calculated on single-CPU computers.

The left panel of Fig. 5.8 shows the spectra for single-ionization into a number of ionic channels labelled by the angular quantum numbers of ion l_i, m_i and the emitted electron l_s. A few-cycle XUV pulse at central photon energy of 58 eV at an (unrealistically high) intensity of 10^{16} W/cm^2 was used, which due to the high photon energy still is in the perturbative regime. The single-photon lowest and first excited ionic channels $(l_i, m_i) = (0, 0)$ dominate ($\sim 58 - 24 = 34$ eV). Due to the high intensity, also the two-photon peaks ($\sim 54 \times 2 - 24 = 84$ eV) are are clearly visible. On the log-plot, the resonant peaks for the excitation of the energetically lowest doubly excited $L = 1$ states of He (2s2p resonance) are well distinct. In the right panel of Fig. 5.8 one sees the build-up rate of strongest of these resonaces in the channel with ionic quantum numbers $(n, l_i, m_i) = (1, 0, 0)$ as a function of time. Note the rather long propagation up to 30 fs. Using streaking techniques [35], those structures may become accessible to experimental observation.

At photon energies of $\sim$120 eV, extremely short XUV pulses containing as few as 2 optical cycles may be produced [36]. Pulse durations are then expected to be comparable or less than the "correlation time" of electrons in the Helium atom. While a precise definition of correlation time is difficult to give, one expects roughly a time scale corresponding to the correlation energy of He $\tau_{corr} \sim \hbar/E_{corr}$. An expected signature is the disappearance of correlation in double electron emission from Helium

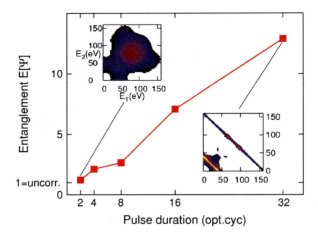

Fig. 5.9 Correlation in the double-emission from Helium as a function of pulse duration at $\hbar\omega = 120\,\text{eV}$. The entanglement measure used is described in [37]. The *insets* show the photo-electron energy dstribution at pulse durations 2 and 32 optical cycles, respectively. For the long pulse, single-photon double emssion ($E_1 + E_2 = 42\,\text{eV}$) and two-photon double emission ($E_1 + E_2 = 162\,\text{eV}$) are clearly distinguishable

in such pulses. Figure 5.9 shows the correlation in the emission part of the wave function as a function of pulse duration. We use the correlation measure discussed, e.g., in [37], where a value of $E[\Psi] = 1$ corresponds an uncorrelated, Hartree-Fock type wave function. The correlation is also reflected in the distribution of photo-electron energies of the two electrons (inserts): at the shortest pulse, photo-electron energies are nearly uncorrelated and each elecron has a broad emission peak near ~ 80 eV. Together with the ground state energy of Helium of 78 eV we approximately obtain the two-photon energy of 240 eV. At the longest pulse, two clearly separated peaks are located at the neutral and ionic single ionization energies $(E_1, E_2) \sim (96 = 120 - 24\,\text{eV}, 66 = 120 - 54\,\text{eV})$ and $(E_1, E_2) \sim (66\,\text{eV}, 96\,\text{eV})$.

5.4 Conclusions and Outlook

With the reduction of the simulation volume afforded by t-SURFF, fully dimensional calculations of double photo-ionization for a broad range of wave-length and intensities come into reach. The exact solution is picked up at some finite surface and knowledge about the long-range behavior of the solutions beyond that surface is exploited. The size reduction is particularly striking for atomic binding potentials with finite range, when the Volkov-solutions become exact at distances where the potential is zero. Electrons that move beyond that range will never scatter and will exactly follow the Volkov solution. For our parameters, the full wave function would expand to several thousand atomic units during the pulse and correspondingly large boxes would be needed, if a spectral analysis of the wave function were performed

5 Photo-Electron Momentum Spectra in Strong Laser-Matter Interactions

after the end of the pulse. In contrast, we could present $\lesssim 1\,\%$ accurate spectra up to energies of 120 eV using a box size of only about 30 atomic units and as little as 75 discretization points per partial wave. With only a few more points, much higher accurcacies can be reached.

Instrumental for the application is the traceless absorption of the wave function beyond the surface, which is provided by the irECS method introduced in [29]. The good performance of irECS for one-dimensional wave functions and for high-harmonic signals from three-dimensional calculations presented in [29] could be confirmed also for the much more delicate observable of angle-resolved photo-electron spectra.

The development of t-SURFF is motivated by the photo-ionization problem. What is special about that problem is that in dipole approximation the time-dependent field affects the whole wave function, including the asymptotic region and therefore final momenta cannot be determined before the end of the pulse, unless one uses knowledge about the time-dependence of the asymptotic solution. However, also in situations without time-dependence of the asymptotic Hamiltonian, the method may turn out to be useful for obtaining fully differential momentum spectra. All it requires are a reliable absorption method and knowledge of the solution in the asymptotic region. Among the candidates for further application are reactive scattering and chemical break-up processes.

References

1. P.B. Corkum, Plasma perspective on strong-field multiphoton ionization. Phys. Rev. Lett. **71**, 1994 (1993)
2. J.L. Chaloupka, J. Rudati, R. Lafon, P. Agostini, K.C. Kulander, L.F. DiMauro, Observation of a transition in the dynamics of strong-field double ionization. Phys. Rev. Lett. **90**, 043003 (2003)
3. M. Meckel, D. Comtois, D. Zeidler, A. Staudte, D. Pavicic, H.C. Bandulet, H. Pepin, J.C. Kieffer, R. Doerner, D.M. Villeneuve, P.B. Corkum, Laser-induced electron tunneling and diffraction. Science **320**(5882), 1478–1482 (2008)
4. M. Spanner, O. Smirnova, P. Corkum, M. Ivanov, Reading diffraction images in strong field ionization of diatomic molecules. J. Phys. B **37**, L243 (2004)
5. S.N. Yurchenko, S. Patchkovskii, I.V. Litvinyuk, P.B. Corkum, G.L. Yudin, Laser-induced interference, focusing, and diffraction of rescattering molecular photoelectrons. Phys. Rev. Lett. **93**, 223003 (2004)
6. F. Martin, J. Fernandez, T. Havermeier, L. Foucar, T. Weber, K. Kreidi, M. Schoeffler, L. Schmidt, T. Jahnke, O. Jagutzki, A. Czasch, E.P. Benis, T. Osipov, A.L. Landers, A. Belkacem, M.H. Prior, H. Schmidt-Boecking, C.L. Cocke, R. Doerner, Single photon-induced symmetry breaking of H-2 dissociation. Science **315**(5812), 629–633 (2007)
7. K. Taylor, J. Parker, K. Meharg, D. Dundas, Laser-driven helium at 780 nm. Eur. Phys. J. D **26**(1), 67–71 (2003)
8. R. Pazourek, J. Feist, S. Nagele, E. Persson, B. Schneider, L. Collins, J. Burgdörfer, Universal features in sequential and nonsequential two-photon double ionization of helium. Phys. Rev. A **83**(5), 1–11 (2011)
9. S. Augst, A. Talebpour, S.L. Chin, Y. Beaudoin, M. Chaker, Nonsequential triple ionization of argon atoms in a high-intensity laser field. Phys. Rev. A **52**, R917–R919 (1995)

10. A. l'Huillier, L.A. Lompre, G. Mainfray, C. Manus, Multiply charged ions induced by multiphoton absorption in rare gases at 0.53 μm. Phys. Rev. A **27**, 2503–2512 (1983)
11. Y. Liu, D. Ye, J. Liu, A. Rudenko, S. Tschuch, M. Dürr, M. Siegel, U. Morgner, Q. Gong, R. Moshammer, J. Ullrich, Multiphoton double ionization of ar and ne close to threshold. Phys. Rev. Lett. **104**, 173002 (2010)
12. R. Moshammer, B. Feuerstein, W. Schmitt, A. Dorn, C.D. Schröter, J. Ullrich, H. Rottke, C. Trump, M. Wittmann, G. Korn, K. Hoffmann, W. Sandner, Momentum distributions of ne^n+ ions created by an intense ultrashort laser pulse. Phys. Rev. Lett. **84**, 447–450 (2000)
13. A. Rudenko, K. Zrost, B. Feuerstein, V.L.B. de Jesus, C.D. Schröter, R. Moshammer, J. Ullrich, Correlated multielectron dynamics in ultrafast laser pulse interactions with atoms. Phys. Rev. Lett. **93**, 253001 (2004)
14. A. Becker, R. Dörner, R. Moshammer, Multiple fragmentation of atoms in femtosecond laser pulses. J. Phys. B: At. Mol. Opt. Phys. **38**(9), S753 (2005)
15. A. Emmanouilidou, Prevalence of different double ionization pathways and traces of three-body interactions in strongly driven helium. Phys. Rev. A **83**, 023403 (2011)
16. N.I. Shvetsov-Shilovski, A.M. Sayler, T. Rathje, G.G. Paulus, Momentum distributions of sequential ionization generated by an intense laser pulse. Phys. Rev. A **83**, 033401 (2011)
17. D.F. Ye, X. Liu, J. Liu, Classical trajectory diagnosis of a fingerlike pattern in the correlated electron momentum distribution in strong field double ionization of helium. Phys. Rev. Lett. **101**, 233003 (2008)
18. M. Awasthi, Y. Vanne, A. Saenz, Non-perturbative solution of the time-dependent Schrodinger equation describing H-2 in intense short laser pulses. J. Phys. B **38**(22), 3973–3985 (2005)
19. H.W. van der Hart, Ionization rates for he, ne, and ar subjected to laser light with wavelengths between 248.6 and 390 nm. Phys. Rev. A **73**, 023417 (2006)
20. J. Parker, B. Doherty, K. Taylor, K. Schultz, C. Blaga, L. DiMauro, High-energy cutoff in the spectrum of strong-field nonsequential double ionization. Phys. Rev. Lett. **96**(13), 7–10 (2006)
21. E. Cormier, P. Lambropoulos, Optimal gauge and gauge invariance in non-perturbative time-dependent calculation of above-threshold ionization. J. Phys. B **29**(9), 1667–1680 (1996)
22. D.A. Telnov, S.I. Chu, Above-threshold-ionization spectra from the core region of a time-dependent wave packet: an ab initio time-dependent approach. Phys. Rev. A **79**(4), 043421 (2009)
23. Z. Zhou, S.I. Chu, Precision calculation of above-threshold multiphoton ionization in intense short-wavelength laser fields: the momentum-space approach and time-dependent generalized pseudospectral method. Phys. Rev. A **83**(1), 33406 (2011)
24. A. Scrinzi, T. Brabec, M. Walser, in 3d numerical calculations of laser-atom interactions, eds. by B. Piraux, K. Rzaszewski. *Proceedings of the Workshop Super-Intense Laser-Atom Physics* (Kluwer Academic Publishers, 2001), p. 313
25. A. Hamido, J. Eiglsperger, J. Madronero, F. Mota-Furtado, P. O'Mahony, A.L. Frapiccini, B. Piraux, Time scaling with efficient time-propagation techniques for atoms and molecules in pulsed radiation fields. Phys. Rev. A **84**(1), 013422 (2011)
26. G.G. Balint-Kurti, R.N. Dixon, C.C. Marston, A.J. Mulholland, The calculation of product quantum state distributions and partial cross-sections in time-dependent molecular collision and photodissociation theory. Comp. Phys. Comm. **63**, 126 (1991)
27. D.J. Tannor, D.E. Weeks, Wave packet correlation function formulation of scattering theory: The quantum analog of classical s-matrix theory. J. Chem. Phys. **98**, 3884 (1993)
28. J. Feist, S. Nagele, R. Pazourek, E. Persson, B.I. Schneider, L.A. Collins, J. Burgdörfer, Non-sequential two-photon double ionization of helium. Phys. Rev. A **77**(4), 043420 (2008)
29. A. Scrinzi, Infinite-range exterior complex scaling as a perfect absorber in time-dependent problems. Phys. Rev. A **81**(5), 053845 (2010)
30. L. Tao, A. Scrinzi, Photo-electron momentum spectra from minimal volumes: the time-dependent surface flux method. New J. Phys. **14**(1), 013021 (2012)
31. O. Koch, W. Kreuzer, A. Scrinzi, Approximation of the time-dependent electronic Schrödinger equation by MCTDHF. Appl. Math. Comput. **173**(2), 960–976 (2006)

32. A.N. Pfeiffer, C. Cirelli, A.S. Landsman, M. Smolarski, D. Dimitrovski, L.B. Madsen, U. Keller, Probing the longitudinal momentum spread of the electron wave packet at the tunnel exit. Phys. Rev. Lett. **109**, 083002 (2012)
33. A. Scrinzi, t-surff: fully differential two-electron photo-emission spectra. New J. Phys. **14**(8), 085008 (2012)
34. E. Goulielmakis, Z.H. Loh, A. Wirth, R. Santra, N. Rohringer, V.S. Yakovlev, S. Zherebtsov, T. Pfeifer, A.M. Azzeer, M.F. Kling, S.R. Leone, F. Krausz, Real-time observation of valence electron motion. Nature **466**(7307), 739 (2010)
35. R. Kienberger, E. Goulielmakis, M. Uiberacker, A. Baltuška, V. Yakovlev, F. Bammer, A. Scrinzi, T. Westerwalbesloh, U. Kleineberg, U. Heinzmann, M. Drescher, F. Krausz, Atomic transient recorder. Nature **427**, 817 (2004)
36. R Kienberger, Private communication
37. R. Grobe, K. Rzazewski, J.H. Eberly, Measure of electron-electron correlation in atomic physics. J. Phys. B: At. Mol. Opt. Phys. **27**(16), L503 (1994)

Chapter 6
Laser Induced Electron Diffraction, LIED, in Circular Polarization Molecular Attosecond Photoionization, MAP

Kai-Jun Yuan and André D. Bandrauk

Abstract We present molecular attosecond photoionization (MAP) with intense ultrashort circularly polarized extreme ultraviolet (XUV) laser pulses by numerically solving appropriate time-dependent Schrödinger equations (TDSEs) of oriented molecules. The resulting molecular photoelectron angular distributions (MPADs) in molecular above threshold ionization (MATI) exhibit signature of laser induced electron diffraction (LIED) with multiple nodes. We adopt a delta function pulse model to describe this effect based on two center interference. Electron *correlation* and *entanglement* in the molecule H_2 at different internuclear distances R due to electron spin exchange are investigated as compared to the delocalized molecular ion H_2^+. For the asymmetric molecular ion HHe^{2+}, LIED patterns can also be observed in MPADs, emphasizing the influence of neighbor Coulomb potentials.

6.1 Introduction

Recent developments in advanced ultrafast intense laser pulse technology [1–3] are catalyzing investigation of electron dynamics in the nonlinear nonperturbative regime of laser-matter interaction. Such ultrashort laser pulse advances have led to the development of imaging pump-probe techniques including laser Coulomb explosion imaging on the femtosecond time scale [4, 5] for nuclear motion to laser induced

---

K.-J. Yuan
Laboratoire de Chimie Théorique, Faculté des Sciences, Université de Sherbrooke, Sherbrooke, QC J1K 2R1, Canada
e-mail: kaijun.yuan@usherbrooke.ca

A. D. Bandrauk (✉)
Canada Research Chair/Computational Chemistry and Molecular Photonics, Laboratoire de Chimie Théorique, Faculté des Sciences, Université de Sherbrooke, Sherbrooke, QC J1K 2R1, Canada
e-mail: andre.bandrauk@USherbrooke.ca

K. Yamanouchi et al. (eds.), *Progress in Ultrafast Intense Laser Science XI*,
Springer Series in Chemical Physics 109, DOI: 10.1007/978-3-319-06731-5_6,
© Springer International Publishing Switzerland 2015

electron diffraction (LIED) [6–12] for coupled electron-nuclear motion on the attosecond (1 as $= 10^{-18}$ s) time scale. Attosecond pulses have now achieved a new ultrashort pulse duration of 67 as [13], thus offering a new tool to watch pure electronic quantum effects without interference from nuclear motion [14], and allowing in particular through attosecond imaging the creation of electron movies [15]. Attosecond pulses allow therefore to separate electronic and nuclear effects due to the different time scales of attosecond for electron motion and femtosecond for nuclear motion.

"Illuminating" molecules from within with modern ultrafast attosecond lasers will allow to use LIED for structural information which is encoded in interference patterns that result from the intrinsic wave nature of the electron and its laser assisted scattering [16]. Direct interference phenomena in diatomics have been predicted earlier by Cohen and Fano [17] and Kaplan and Markin [18] in perturbative single photon ionization. This has been extended to nonperturbative photoionization with few-cycle intense laser pulses in LIED [6, 9]. By taking the ratio of the momentum distributions in the two lateral directions, diffractive interference and diffraction for oriented molecules can be obtained [19]. In these schemes, a laser induced *rescattering* mechanism occurs through recombination with the parent ion [20, 21] or with neighboring ions [22, 23] with electron wavelengths controlled by the laser intensity through laser induced ponderomotive energies. The effects of orbital geometry on tunnelling-rescattering ionization and the resulting interference have also been investigated in angular resolved high order molecular above threshold ionization (MATI) spectra [24]. This has been extended to circularly polarized pulses illustrating molecular photoelectron angular distributions (MPADs) in MATI [25] and molecular high-order harmonic generation (MHOHG) [26–28] both at equilibrium and at large internuclear distance R. On an attosecond time scale a polyatomic molecule CO_2 was simulated for imaging geometric and orbital structures using LIED techniques excited by a near-infrared few-cycle laser pulse [10]. It is found that the molecular geometry (bond lengths) is determined with 3 % accuracy from the corresponding diffraction pattern which also reflects the nodal properties of the initial molecular orbital. Recent investigations have shown that classical diffraction is only appropriate at substantially higher photon energies [29]. In these processes the MPAD is defined the direct transition amplitude from the initial states to the continuum. The coherent emission of the spherical electron wave from the two spatially separated centers gives the interference patterns. The effects of the spherical wave in the molecular continuum on electron diffraction and interference has also been investigated within the framework of a zero-range potential model [30]. Double-slit interference effects in the vibrational resolved cross section of electron emission from diatomic molecules were recently revealed [31, 32].

Circularly polarized attosecond laser pulses have now been used as a new powerful tool for studying electron dynamics in atoms, molecules and solid materials [33–47]. We have proposed previously methods of generating circularly polarized attosecond pulses from circularly polarized MHOHG [48, 49] using the nonspherical symmetry of molecules as opposed to spherical symmetry in atoms which prohibits circularly polarized harmonics. Using two color linearly polarized orthogonal laser

6 Laser Induced Electron Diffraction

pulses it has been found that circularly polarized attosecond pulses can also be produced in oriented molecules [50]. Circularly polarized laser pulses are shown to be important to induce electron ring current in cyclic molecules, thus producing intense internal magnetic fields [51, 52] and to investigate recollision of electrons in double ionization [53, 54]. Circular laser polarization has been emphasized earlier to be important in strong field ionization processes [20]. By intense few cycle circularly polarized XUV pulses, strong attosecond magnetic field pulses with several tens of Teslas can be induced in molecules [55]. In this review chapter, we present molecular attosecond photoionization (MAP) with circularly polarized XUV laser pulses by numerically solving the time dependent Schrödinger equation (TDSE). MPADs exhibit LIED patterns in molecules H_2, H_2^+, and HHe^{2+}, which are shown to be sensitive to the symmetry of the electronic state, molecular internuclear distance R and pulse wavelength λ. Nuclear motion is not discernible (measurable) due to the large bandwidth and duration shorter than the femtosecond time scale [56, 57]. We focus therefore on MAP with static nuclei and show effects of electron diffraction in MPADs.

The chapter is organized as follows: We first present the numerical method for simulations of MAP in Sect. 6.2. LIED patterns in the two electron H_2 molecule and the single electron H_2^+ molecular ion are illustrated in Sect. 6.3. The results for the asymmetric molecular ion HHe^{2+} are shown in Sect. 6.4. Dependence of MATI spectra on the pulse wavelength λ is investigated in Sect. 6.5. Finally, we summarize our finding in Sect. 6.6. Throughout this paper, atomic units (a.u.) $e = \hbar = m_e = 1$ are used.

6.2 Numerical Methods

The TDSE for a fixed nuclei pre-oriented two electron molecule H_2 is appropriately written with respect to the center of mass of two protons at position $(\pm R/2, 0, 0)$ as,

$$i\frac{\partial}{\partial t}\psi(t) = \left[-\frac{1}{2}\nabla_1^2 - \frac{1}{2}\nabla_2^2 + V_{ee} + V_{en} + V_L(t)\right]\psi(t), \tag{6.1}$$

where $\nabla_{1,2}^2$ is the Laplacian of each electron on the right hand side of (6.1), V_{ee} is the electron-electron repulsion and the electron-proton attraction Coulomb potentials is V_{en}. The TDSE is described using cylindrical coordinates (ρ, θ, z) with $(x = \rho \cos\theta, y = \rho \sin\theta)$, where the radial and angular variables are separated from z, i.e., the z axis is perpendicular to the molecular (x, y) plane, as illustrated in Fig. 6.1. R is the molecular distance aligned along the x axis. We reduce the 3D problem to 2D by separating the perpendicular z-coordinate. 2D calculations are well suited for circularly polarized laser pulses because such a pulse confines electrons in the plane of laser polarization. V_{ee} is the electron-electron repulsion, and the electron-proton attraction Coulomb potentials is V_{en},

Fig. 6.1 Schematic illustrations of photoionization processes for the x aligned molecular ion H_2^+ by a circularly polarized XUV laser pulse $E(t)$. The ionizing laser pulse is polarized in the molecular (x, y) plane, propagating along the perpendicular z axis

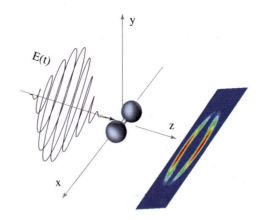

$$V_{ee} = \left\{ \rho_1^2 + \rho_2^2 - 2\rho_1\rho_2 \cos(\theta_1 - \theta_2) + \alpha \right\}^{-1/2}, \quad (6.2)$$

and

$$V_{en} = -\sum_{j=1}^{2} \left\{ \rho_j^2 + R\rho_j \cos\theta_j + R^2/4 + \beta \right\}^{-1/2}, \quad (6.3)$$

α and β are regularization parameters to remove Coulomb singularities (this corresponds to averaging over the third dimension perpendicular to the plane of the molecule, i.e., the z direction) The regularization parameters are chosen to be $\alpha = 1.0$ and $\beta = 0.35$, allowing the accurate reproduction of the ground and excited potential energies of H_2 and H_2^+.

The laser-electron radiative coupling is described by the time dependent potential

$$V_L(t) = -E_0 f(t) \left[\sum_{j=1}^{2} \rho_j \left(\cos\theta_j \cos\omega t + \sin\theta_j \sin\omega t \right) \right], \quad (6.4)$$

for an $E(t)$ field in the molecular (x, y) plane. A temporal slowly varying envelope $f(t) = \sin^2(\pi t/T)$ with duration $T = n_l \tau$ (1 optical cycle, o.c., $\tau = 2\pi/\omega$) and maximum amplitude E_0 for intensity $I_0 = \frac{1}{2} c \varepsilon_0 E_0^2$ is adopted, so that the total pulse area $A(t) = -\frac{1}{c} \int_{t_i}^{t_f} E(t) dt$, where t_i is any time before the start of the pulse envelope and t_f is any time after the turn off of the pulse envelope, and $A(t_f) - A(t_i) = 0$ which is constrained in accordance with Maxwell's equation [1].

The TDSE is solved numerically by a second-order split-operator method in the time step δt combined with a fifth order finite difference method and Fourier transform technique in the spatial steps $\delta\rho$ and $\delta\theta$ [58, 59]. The time step is taken to be $\delta t = 0.01$ a.u. = 0.24 as, the spatial discretization is $\delta\rho = 0.25$ a.u. for radial grid range $0 \leq \rho \leq 128$ a.u. and the angle grid size $\delta\theta = 0.025$ radian. To prevent unphysical effects due to the reflection of the wave packet from the boundary, we

6 Laser Induced Electron Diffraction

multiply $\psi(\rho, \theta, t)$ by a "mask function" or absorber in the radial coordinate with the form

$$\mathscr{G}(t) = \begin{cases} 1, & \rho < \rho_a \\ \cos^{1/8}[\pi(\rho - \rho_a)/2\rho_{abs}], & \rho_a \leq \rho \leq \rho_{max} \end{cases}. \tag{6.5}$$

For all results reported here we set the absorber domain $\rho_a = \rho_{max} - \rho_{abs} = 104$ a.u. with $\rho_{abs} = 24$ a.u., exceeding well the field induced electron oscillation $\alpha_d = E_0/\omega^2$ of the electron.

An energy analysis of the ionization probabilities is employed based on a Fourier analysis of the associated flux (electronic current density) to describe MPADs [60, 61], which has been proposed to simulate momentum distributions of nuclear and electronic wave packets [62, 63]. For attosecond photoionization processes, the laser duration T is very short. To eliminate quiver effects of ionized electrons, a large asymptotic point is required. At such large asymptotic point ρ_0, the angular flux distributions can be ignore, i.e., $1/\rho_0 \partial/\partial\theta|_{\rho_0}\hat{e}_\theta \ll \partial/\partial\rho|_{\rho_0}\hat{e}_\rho$. The time-independent energy-resolved angular differential yield (photoelectron spectra) is obtained by a Fourier transform:

$$\Psi(\theta, E_e)|_{\rho_0} = \int_{t_p}^{\infty} \psi(\theta, t)|_{\rho_0} e^{iE_e t} dt,$$

$$\Psi'(\theta, E_e)|_{\rho_0} = \int_{t_p}^{\infty} \frac{\partial \psi(\theta, t)}{\partial \rho}|_{\rho_0} e^{iE_e t} dt, \tag{6.6}$$

$$\mathscr{J}(\theta, E_e) \sim \Re\left[\frac{1}{2i}\Psi'^*(\theta, E_e)|_{\rho_0}\Psi(\theta, E_e)|_{\rho_0}\right],$$

where t_p is time after pulses switch off. The radial flux is calculated at $\rho_0 = 100$ a.u.. $E_e = p_e^2/2$ is the kinetic energy of an ionized electron with wave vector $k = p_e = 2\pi/\lambda_e$ (in a.u.), $p_e = (p_x^2 + p_y^2)^{1/2}$ is the momentum of a photoelectron of wavelength λ_e.

Since only the radial electronic flux is taken into account in simulations, we may define $\theta_{pR} = \theta$ where θ_{pR} is the angle between the electron momentum $\mathbf{p}$ and the molecular axis $\mathbf{R}$. With the transformation $p_x = p_e \cos\theta_{pR}$ and $p_y = p_e \sin\theta_{pR}$, we then obtain the 2D momentum distributions of a photoelectron from (6.6). The total MPAD $\mathscr{J}^{total}(\theta_{pR})$ can be obtained by integrating over energy,

$$\mathscr{J}^{total}(\theta_{pR}) = \int_0^{\infty} dE_e \mathscr{J}(\theta_{pR}, E_e). \tag{6.7}$$

The MPAD at a specific kinetic energy E_{en} is obtained by integrating over energy,

$$\mathcal{J}^{E_{en}}(\theta_{pR}) = \int_{E_{en}-\omega/2}^{E_{en}+\omega/2} dE_e \, \mathcal{J}(\theta_{pR}, E_e). \tag{6.8}$$

For two-electron molecules H_2, the corresponding angular distributions are reduced to single-electron spectra by integrating out one electron.

6.3 Diffraction in H_2 and H_2^+

The H_2 molecule and its isotopomers, HD and D_2, have been used as a benchmark system of the two-electron chemical bond with bonding and antibonding molecular orbitals the essential concepts for understanding the chemical bond [64]. We will focus on the first three electronic states of H_2, $X^1\Sigma_g^+$, $A^3\Sigma_u^+$, and $B^1\Sigma_u^+$ with the following molecular orbital configurations at "equilibrium" $R_e = 1.675$ a.u. (numerical values) [60, 61] and their valence bond (Heitler-London) atomic configurations [64] at large internuclear distance $R = 10$ a.u.;

$$X^1\Sigma_g^+ : R = R_e, \quad 1\sigma_g(1)1\sigma_g(2),$$
$$R = 10\,a.u., \quad 1s_a(1)1s_b(2) + 1s_a(2)1s_b(1), \tag{6.9}$$

$$A^3\Sigma_u^+ : R = R_e, \quad 1\sigma_g(1)1\sigma_u(2) - 1\sigma_g(2)1\sigma_u(1),$$
$$R = 10\,a.u., \quad 1s_a(1)1s_b(2) - 1s_a(2)1s_b(1), \tag{6.10}$$

$$B^1\Sigma_u^+ : R = R_e, \quad 1\sigma_g(1)1\sigma_u(2) + 1\sigma_g(2)1\sigma_u(1),$$
$$R = 10\,a.u., \quad 1s_a(1)1s_a(2) - 1s_b(1)1s_b(2), \tag{6.11}$$

where $1s_a$ and $1s_b$ denote the $1s$−like atomic orbitals on protons a and b, respectively, and 1 and 2 are the coordinates of the two electrons. Both $X^1\Sigma_g^+$ and $A^3\Sigma_u^+$ states treated in dissociate at large R to ground state H(1s) atoms but with different spin symmetry leading to electron entanglement via exchange [65]. The $B^1\Sigma_u^+$ state dissociates to the ionic fragments H^+H^- as shown by extensive multiconfiguration interaction (CI) calculations [66]. This is shown in Fig. 6.2 where we have projected in one-dimension (1D), i.e., in electron coordinates x_1 and x_2, the $X^1\Sigma_g^+$ wavefunction in Fig. 6.2a, the $A^3\Sigma_u^+$ wavefunction in Fig. 6.2b and the $B^1\Sigma_u^+$ wavefunction in Fig. 6.2c at $R = 10$ a.u.. Here polar coordinates (ρ, θ) are used for describing electron motions, where $x = \rho \cos \theta$ and $y = \rho \sin \theta$, and the molecular axis is chosen as the x−axis in Fig. 6.2. In the $X^1\Sigma_g^+$ and $A^3\Sigma_u^+$ states, electrons localize mainly at $x_1 = -x_2$ on separate nuclei, whereas in the $B^1\Sigma_u^+$ state at $x_1 = x_2$, i.e., on the same atom, corresponding to the ionic states H^+H^- and H^-H^+.

6 Laser Induced Electron Diffraction

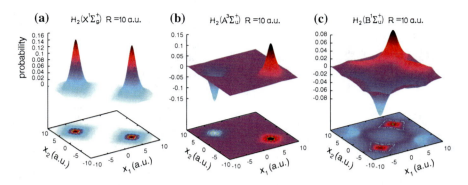

Fig. 6.2 Amplitudes of the two-electron wavefunctions of H_2 at internuclear distance $R = 10$ a.u. for **a** the $X^1\Sigma_g^+$ state where electrons localize at $x_1 = -x_2$, **b** the $A^3\Sigma_u^+$ state, where electrons mainly localize at $x_1 = -x_2$ but out of phase, and **c** the $B^1\Sigma_u^+$ state where electrons mainly localize at $x_1 = x_2 = -R/2$ and $x_1 = x_2 = R/2$, obtained from $\mathscr{P}(x_1, x_2) = \int\int \psi(x_1, y_1, x_2, y_2) dy_1 dy_2$

With intense ultrashort pulses these ionic states cross the initial covalent ground state and are responsible for the unique phenomenon of EI at certain large critical internuclear distances [67, 68]. We compare the calculated H_2 MPADs for linearly and circularly polarized XUV laser pulses with the H_2^+ $X^2\Sigma_g^+$ $(1\sigma_g)$ and $A^2\Sigma_u^+$ $(1\sigma_u)$ electronic states. As emphasized in [69, 70], the correlated interaction of two electrons in H_2 can significantly distort the interference patterns that is expected when a single electron, as in H_2^+, is diffracted by two nuclei. In the present work we examine such effects in the first three electronic states of H_2, $X^1\Sigma_g^+$, $A^3\Sigma_u^+$, and $B^1\Sigma_u^+$ at "equilibrium" and large internuclear distances. Therefore, at "equilibrium" R_e, electron configurations are well described by independent particle molecular orbitals, $1\sigma_g$ and $1\sigma_u$ thus involving little correlation. At large distance, as seen from (6.9), the appropriate configurations are "entangled" atomic orbital states. Spin exchange entanglement is now considered an important tool for quantum computing as in coupled quantum dots [65].

6.3.1 MAPDs in H_2 and H_2^+ by XUV Pulses

We consider the single photon circular polarization ionization, i.e., MPADs in the first order MATI at energy $\omega - I_p$. Table 6.1 shows the parameters used in the simulations at different internuclear distances R, laser wavelength λ (angular frequency $\omega = 2\pi c/\lambda$), corresponding ionized electron de Broglie wavelength λ_e and momentum p obtained from $p\lambda_e = h = 2\pi$ (in a.u.) and the energy conservation (in a.u.) $\omega = I_p + 2\pi^2/\lambda_e^2$, where I_p is the ionization potential of each electronic state. Figures 6.3, 6.4, 6.5, 6.6, 6.7, 6.8, 6.9 and 6.10 show the corresponding MPADs $\mathscr{J}(\theta)$. We note that at the H_2 "equilibrium", $R_e = 1.675$ a.u., λ_e ranges from approximately $3R_e/2$ to R_e whereas at the extended distance $R = 10$ a.u. $\lambda_e \approx R/4$

Table 6.1 Wavelength λ_e (momentum $p = 2\pi/\lambda_e$ in a.u.) of the photoelectron in H_2 and H_2^+ ionization at wavelengths λ of the ionizing laser pulse obtained by $\omega(\text{a.u.}) \sim I_p + 2\pi^2/\lambda_e^2$, where I_p is the vertical ionization potential

	$R_e = 1.675$ a.u.		$R = 10$ a.u.	
	$\lambda = 10$ nm	$\lambda = 5$ nm	$\lambda = 10$ nm	$\lambda = 5$ nm
	λ_e (p)	λ_e (p)	λ_e (p)	λ_e (p)
$X^1\Sigma_g^+$ (H_2)	2.75 (2.28) a.u.	1.65 (3.81) a.u.	2.44 (2.58) a.u.	1.59 (3.95) a.u.
$A^3\Sigma_u^+$ (H_2)	2.58 (2.44) a.u.	1.62 (3.88) a.u.	2.44 (2.58) a.u.	1.59 (3.95) a.u.
$B^1\Sigma_u^+$ (H_2)	2.50 (2.51) a.u.	1.61 (3.90) a.u.	2.30 (2.73) a.u.	1.55 (4.05) a.u.
$X^2\Sigma_g^+$ (H_2^+)	2.45 (2.56) a.u.	1.58 (3.98) a.u.	2.26 (2.78) a.u.	1.53 (4.10) a.u.
$A^2\Sigma_u^+$ (H_2^+)	2.27 (2.77) a.u.	1.54 (4.08) a.u.	2.26 (2.78) a.u.	1.53 (4.10) a.u.

and $R/6$ for $\lambda = 10$ and 5 nm. In this large distance configuration one should therefore expect considerable LIED due to electron-proton scattering. Increasing the internuclear distance R of the molecules by fixing the photon energy is similar to the concept of increasing the photon energy to match the internuclear distance [29]. The effects of electron correlation and entanglement on different electronic states, atomic configurations, are dependent on the internuclear distance R as well, resulting in distribution characteristic of photoionization.

Single electron photoionization leads generally to simple expressions for the MPADs. In the case of H_2^+ and H_2 initial Σ electronic states considered here, with initial zero angular momentum, the photoionized electron is expected a dipole $\cos^2 \theta$ MPAD for linear parallel polarization and a $\sin^2 \theta$ distribution for linear perpendicular polarization. This simple ionization pattern is clearly reproduced in Figs. 6.3 and 6.4, for linear polarizations, except for the H_2^+ $A^2\Sigma_u^+$ distributions in Fig. 6.4, where both linear parallel and perpendicular ionizations are anomalous and different than in H_2. Furthermore Figs. 6.3 and 6.4 for circular polarizations appear as incoherent, i.e., additive, superpositions of both parallel and perpendicular linear polarization ionizations. From Table 6.1, in Figs. 6.3 and 6.4 for the photon wavelength $\lambda = 10$ nm and $R_e = 1.675$ a.u., the electron wavelength $\lambda_e > R_e$. Under such condition no electron diffraction is therefore obtained in total MPADs in Figs. 6.3 and 6.4. Electron diffraction effects begin to occur for $\lambda_e < R$ in Figs. 6.5, 6.6, 6.7, 6.8, 6.9 and 6.10 resulting in more complex total MPADs. The signature of these diffraction effects appears clearly as nodes in the corresponding MPADs. The slightly nonsymmetric circular polarization angular distributions in Fig. 6.5 for both H_2 and H_2^+ Σ_g states and in Fig. 6.6 for the H_2^+ $A^2\Sigma_u^+$ state can be attributed to incipient coherent interference between Σ and Π electronic transitions [71]. Of note is that the H_2 $A^3\Sigma_u^+$ and $B^1\Sigma_u^+$ states show different MPADs, as opposed again to the H_2^+ $A^2\Sigma_u^+$ state in Fig. 6.6 for all three electronic states. Molecular orbital theory, (6.9) predicts that the ionized electron emanates from a same antisymmetry orbital, $1\sigma_u \simeq (1s_a - 1s_b)/\sqrt{2}$ in these three electronic states. Clearly as seen from the MPADs, the H_2 $A^3\Sigma_u^+$ and H_2^+ $A^2\Sigma_u^+$ states in Fig. 6.6c and i, these two states have six nodes whereas four nodes appear in the H_2 $B^1\Sigma_u^+$ state in Fig. 6.6f. The latter

$H_2(X^1\Sigma_g^+)$ R_e=1.675 a.u. λ_e=2.75 a.u.

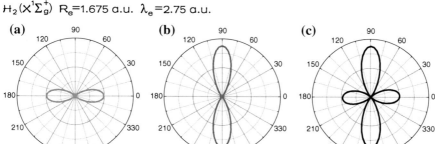

$H_2^+(X^2\Sigma_g^+)$ R_e=1.675 a.u. λ_e=2.45 a.u.

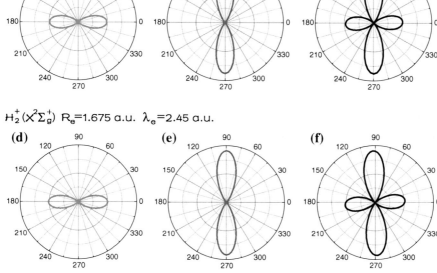

Fig. 6.3 MPADs for (**a–c**, *top row*) the H_2 $X^1\Sigma_g^+$ and (**d–f**, *bottom row*) the H_2^+ $X^2\Sigma_g^+$ ground electronic states at $\lambda = 10$ nm, $R_e = 1.675$ a.u., $\lambda_e > R_e$, for (**a, d**, *left column*) parallel *linear* polarization, (**b, e**, *middle column*) perpendicular *linear* polarization, and (**c, f**, *right column*) *circular* polarization ionizations

in fact shows the appearance of a maximum at 90° and 270° as opposed to the two other electronic states.

Figures 6.7, 6.8, 6.9 and 6.10 correspond to the ionized electron wavelengths $\lambda_e \simeq R/4$ and $R/6$ (Table 6.1) for photoionization with a photon wavelength $\lambda = 10$ and 5 nm at $R = 10$ a.u.. Similar processes occur for both cases. Comparing the H_2^+ ground state $X^2\Sigma_g^+$ in Figs. 6.7f and 6.9f, and the first excited state $A^2\Sigma_u^+$ in Figs. 6.8i and 6.10i MPADs shows same signatures due to the highly delocalized nature of the electron over the two separated protons. The orbital symmetry affects slightly the intensities of emerging electron "jets" in H_2^+ at the same angles for linear parallel and perpendicular ionizations, i.e, Fig. 6.7d, e versus Fig. 6.8g, h and Fig. 6.9d, e versus Fig. 6.10g, h. The H_2 $X^1\Sigma_g^+$ MPADs are quite different from the corresponding H_2^+ $X^2\Sigma_g^+$ results in Figs. 6.7a–c and 6.9a–c. This reflects the difference in electronic structure since the ground state of H_2 at $R = 10$ a.u. is essentially the atomic Heitler London entangled electronic configurations $[1s_a(1)1s_b(2) + 1s_a(2)1s_b(1)]$, (6.9) as compared to the completely delocalized $1\sigma_g \sim (1s_a + 1s_b)/\sqrt{2}$ orbital of

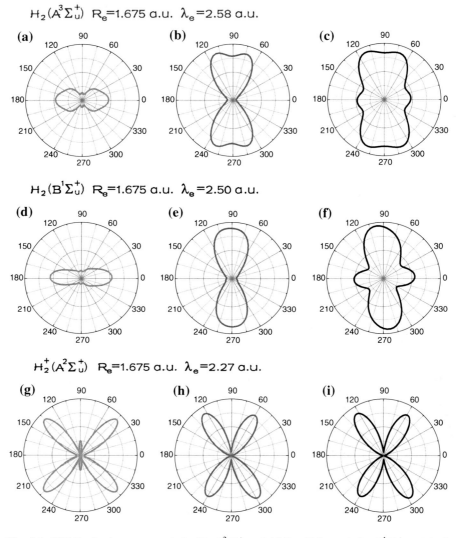

Fig. 6.4 MPADs for (**a–c**, *top row*) the H_2 $A^3\Sigma_u^+$ and (**d–f**, *middle row*) the $B^1\Sigma_u^+$ and (**g–i**, *bottom row*) the H_2^+ $A^2\Sigma_u^+$ excited electronic states at $\lambda = 10$ nm, $R_e = 1.675$ a.u., $\lambda_e > R_e$, for (**a, d, g**, *left column*) parallel *linear* polarization, (**b, e, h**, *middle column*) perpendicular *linear* polarization, and (**c, f, i**, *right column*) *circular* polarization ionizations

H_2^+. Surprisingly the H_2 $B^1\Sigma_u^+$ state shows similar angular distributions as the H_2^+ $A^2\Sigma_u^+$ state in Figs. 6.8 and 6.10 whereas the H_2 $A^3\Sigma_u^+$ state is more similar to the ground $X^1\Sigma_g^+$ state—This reflects the near degeneracy at large R of these two latter electronic states, where they both evolve into entangled Heitler London atomic electronic configurations, (6.9). For the $X^1\Sigma_g^+$ and $A^3\Sigma_u^+$ states of H_2 in circular

6 Laser Induced Electron Diffraction

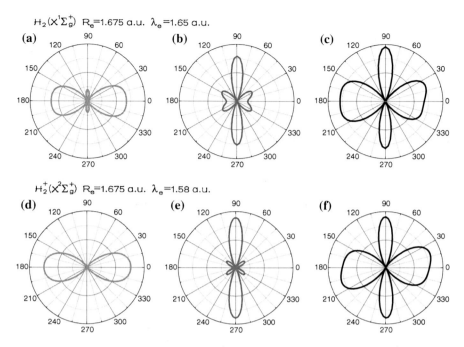

Fig. 6.5 MPADs for (**a**–**c**, *top row*) the H_2 $X^1\Sigma_g^+$ and (**d**–**f**, *bottom row*) the H_2^+ $X^2\Sigma_g^+$ ground electronic states at photon wavelength $\lambda = 5$ nm, $R_e = 1.675$ a.u., $\lambda_e \simeq R_e$, for (**a**, **d**, *left column*) parallel *linear* polarization, (**b**, **e**, *middle column*) perpendicular *linear* polarization, and (**c**, **f**, *right column*) *circular* polarization ionizations

polarization ionization i.e., Fig. 6.7c versus Fig. 6.8c and Fig. 6.9c versus Fig. 6.10c, more diffraction is observed for the $A^3\Sigma_u^+$ state.

6.3.2 Description of LIED in MPADs

Electron diffraction effects appear for $\lambda_e < R$ in Figs. 6.7 and 6.8, as distortions of the single atomic dipole angular distributions for $\lambda_e > R$ in Figs. 6.3 and 6.4, with the same photon wavelength $\lambda = 10$ nm. In particular, for the case of strong diffraction effects, i.e., for $\lambda_e < R$, strong electron "jets" appear for H_2^+ $X^2\Sigma_g^+$ and $A^2\Sigma_u^+$ states for all polarizations, linear and circular. Such ejections of electron "jets" dominate also in the H_2 $B^1\Sigma_u^+$ state, with similar structure as the H_2^+ $X^2\Sigma_g^+$ and $A^2\Sigma_u^+$ states in Fig. 6.8. To explain the occurrence of such electron "jets" we use the exactly solvable photoionization model of H_2^+ by a delta function pulse in LIED [6], shown in Appendix 1. Using the expression in (6.16) and (6.17) we see indeed that in Figs. 6.3 and 6.4, and Table 6.1, the condition $\cos\theta_{pR} > 1$, but that such θ_{pR} does not exit, where θ_{pR} is the angle between the electron momentum **p**

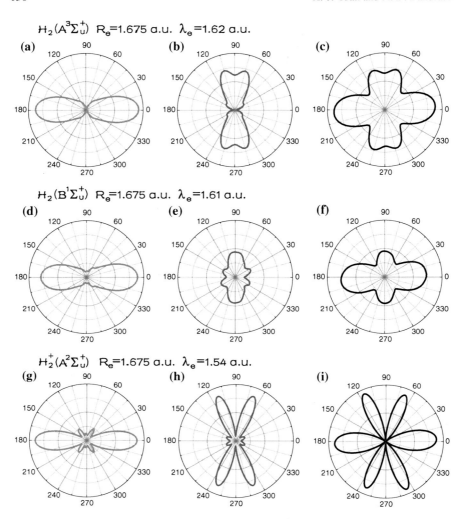

Fig. 6.6 MPADs for (**a–c**, *top row*) the H_2 $A^3 \Sigma_u^+$ and (**d–f**, *middle row*) the $B^1 \Sigma_u^+$ and (**g–i**, *bottom row*) the H_2^+ $A^2 \Sigma_u^+$ excited electronic states at $\lambda = 5$ nm, $R_e = 1.675$ a.u., $\lambda_e \simeq R_e$, for (**a, d, g**, *left column*) parallel *linear* polarization, (**b, e, h**, *middle column*) perpendicular *linear* polarization, and (**c, f, i**, *right column*) *circular* polarization ionizations

and the orientation of the **R**—internuclear axis. Then no diffraction is possible and the MPADs for both Σ_g and Σ_u states in H_2 and H_2^+ are essentially the same and are atomic. The first indication of diffraction appears in Figs. 6.5 and 6.6 at $\lambda = 5$ nm and $R = 1.675$ a.u.. Both Σ_g states in H_2 and H_2^+ show similar MPADs in Fig. 6.5. The same behavior is observed in Fig. 6.6 for the Σ_u states of H_2 and H_2^+. For the Σ_g states, the ionized electron is ionized from a symmetry like $1\sigma_g$ orbital whereas for the Σ_u states the ionized orbital is $1\sigma_u$ in both H_2 and H_2^+. In Figs. 6.5 and 6.6,

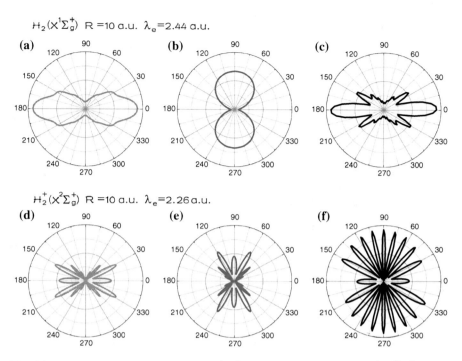

Fig. 6.7 MPADs for (**a–c**, *top row*) the H$_2$ $X^1\Sigma_g^+$ and (**d–f**, *bottom row*) the H$_2^+$ $X^2\Sigma_g^+$ ground electronic states at $\lambda = 10$ nm, $R = 10$ a.u., $\lambda_e < R$, for (**a, d**, *left column*) parallel *linear* polarization, (**b, e**, *middle column*) perpendicular *linear* polarization, and (**c, f**, *right column*) *circular* polarization ionizations

Σ_g and Σ_u states show a dipole $\cos\theta_e$ atomic like distribution for linear parallel ionization. Due to the fact that $\cos\theta_{pR} > 1$ as in Figs. 6.3 and 6.4, i.e., no diffraction is possible in these cases. Σ_u states in both H$_2$ and H$_2^+$ show perpendicular ionization diffraction for electrons ionizing out of $1\sigma_u$ orbitals in Fig. 6.6. Small "wings" occur at $\theta = 30°$ in Fig. 6.5 and near 60° in Fig. 6.6 for the Σ_g^+ and Σ_u^+ states of H$_2$ and H$_2^+$ at perpendicular linear ionization. We note that already for the H$_2^+$ $A^2\Sigma_u^+$ state at $\lambda = 10$ nm, $R_e = 1.675$ a.u., the $\sin(pR\cos\theta_{pR}/2)$ term is maximum at $\theta_{pR} = 45°$ since $pR/2 = \pi R/\lambda_e = 0.7\pi$ (Table 6.1). This explains the anomalous MPADs for that state in Fig. 6.4. At $\lambda = 5$ nm in Fig. 6.6, $\pi R/\lambda_e = \pi$ for the H$_2^+$ $A^2\Sigma_u^+$ state. Then $\sin(\pi\cos\theta_{pR})$ is maximum at $\pi\cos\theta_{pR} = \pi/2$ or $\theta_{pR} = 60°$. Small "wings" do occur for the linear parallel ionization in Fig. 6.6 and strong ionization at 60° and 120° for the linear perpendicular case in Fig. 6.6. We note that the H$_2$ $B^1\Sigma_u^+$ state has strong linear perpendicular ionization in Fig. 6.6 thus suggesting that the initial $1\sigma_g$ orbital is contributing to the ionization.

Figures 6.7 and 6.8 illustrate much more complex angular distributions for the momentum of ionized electrons in H$_2$ and H$_2^+$, thus reflecting dominant diffraction effects since $\lambda_e < R$. Firstly one notices that the ground $X^1\Sigma_g^+$ and the excited

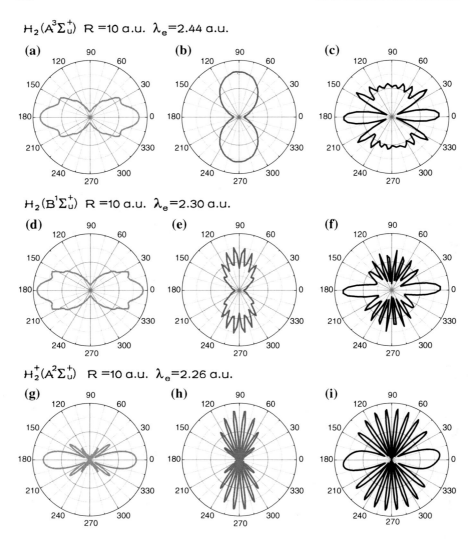

Fig. 6.8 MPADs for (**a**–**c**, *top row*) the H_2 $A^3\Sigma_u^+$ and (**d**–**f**, *middle row*) the $B^1\Sigma_u^+$ and (**g**–**i**, *bottom row*) the H_2^+ $A^2\Sigma_u^+$ excited electronic states at $\lambda = 10$ nm, $R = 10$ a.u., $\lambda_e < R$, for (**a, d, g**, *left column*) parallel *linear* polarization, (**b, e, h**, *middle column*) perpendicular *linear* polarization, and (**c, f, i**, *right column*) *circular* polarization ionizations

$A^3\Sigma_u^+$ states of H_2 at $R = 10$ a.u. for $\lambda = 10$ nm show very similar distributions for linear and circular polarizations due to the fact both states dissociate at this large internuclear distance to the same Heitler London atomic orbitals, (6.9) and (6.10). The only difference in the total two-electron wavefunctions is the symmetric (6.9) versus antisymmetric (6.10) and Fig. 6.2 entanglement via exchange. Both parallel and perpendicular polarization ionizations are identical. The circular polarization

6 Laser Induced Electron Diffraction

distributions in Figs. 6.7 and 6.8, are also nearly identical with maxima occurring at $\theta_e = 0°, 30°$ and $60°$, which are not direct (incoherent) superposition of linear parallel in Figs. 6.7 and 6.8, and perpendicular Figs. 6.7 and 6.8 ionizations. None of the H_2 $X^1\Sigma_g^+$ and $A^3\Sigma_u^+$ ionization distributions resemble the corresponding H_2^+ distributions in Figs. 6.7 and 6.8. This reflects completely the difference in electronic configurations, entangled atomic states, (6.9) and (6.10) for H_2 vs a delocalized single electronic state for H_2^+. The extended delocalized nature of the electron in $1\sigma_g \simeq (1s_a + 1s_b)/\sqrt{2}$ and $1s\sigma_u \simeq (1s_a - 1s_b)/\sqrt{2}$ molecular orbitals shows nearly identical high interference patterns in ionizations for the H_2^+ $X^2\Sigma_g^+$ (Fig. 6.7) and $A^2\Sigma_u^+$ (Fig. 6.8) states. In both cases the linear parallel (Fig. 6.7) and perpendicular (Fig. 6.7) distributions are a rotation by $\pi/2$ reflecting the dominance of the atomic distributions with multiple electron "jets" at different angles θ_e due to diffraction. The circular polarization distributions in Figs. 6.7 and 6.8 are identical for the H_2^+ $X^2\Sigma_g^+$, $A^2\Sigma_u^+$ and H_2 $B^1\Sigma_u^+$ states as linear superpositions of comparable parallel and perpendicular linear polarization ionizations. For parallel ionization, using $p = 2.5$ a.u. at $R = 10$ a.u., the interference term $\cos(\mathbf{p} \cdot \mathbf{R}/2) = \cos(pR\cos\theta_{pR}/2)$ has maxima at $\cos\theta_{pR} \simeq n/4 = 1/4, 1/2, 3/4$, and 1, whereas for perpendicular ionization, $\sin(\mathbf{p} \cdot \mathbf{R}/2)$ has maxima at $\cos\theta_{pR} = (2n+1)/8 = 1/8, 3/8, 5/8$, and $7/8$. Figure 6.8 for the H_2 $B^1\Sigma_u^+$ and H_2^+ $A^2\Sigma_u^+$ states as well as Fig. 6.7 for the H_2^+ ground $X^2\Sigma_g^+$ state show more structures than the simple predictions from (6.17). This can be explained as coming from the extreme nonsphericity of H_2 and H_2^+ at large distances [71, 72] The most surprising result is the near identical angular ionization distributions for the H_2 $B^1\Sigma_u^+$ and H_2^+ $A^2\Sigma_u^+$ states in Fig. 6.8 for all polarizations. Inspection of the form of the electronic configurations, (6.10) versus (6.11), shows that both of these H_2 and H_2^+ states involve one electron delocalized over two centers. For the $B^1\Sigma_u^+$ state a $H^+H^- + H^-H^+$ configuration is a loosely bound pair of $1s$ electrons, H^- localized and delocalized by entanglement and exchange, whereas in the H_2^+ $A^2\Sigma_u^+$ state it is a $1s$ electron delocalized by its wave nature. Figure 6.2 for the $B^1\Sigma_u^+$ state wavefunction shows both localization into H^-H^+ states along the line $x_1 = x_2$ and also considerable delocalization at $x_1 = -x_2$. The single electron MPAD does not distinguish clearly between exchange entanglement and pure quantum delocalization, due to the identical antisymmetry of the wavefunctions in both cases. The same behaviors are produced for $\lambda = 5$ nm at $R = 10$ a.u., Figs. 6.9 and 6.10, but more electron "jets" are emitted in MPADs, as predicted by (6.16) and (6.17) in Appendix 1.

6.4 LIED in Asymmetric HHe^{2+}

In Sect. 6.3 we showed results of symmetric molecules H_2 and H_2^+. Under conditions $\lambda_e < R$, LIED occurs. For asymmetric molecules we find that LIED patterns can also be observed in MPADs. The single electron molecular ion HHe^{2+} serves as the simplest asymmetric model which has revealed many novel phenomena including multichannel MHOHG [73, 74] via excited state resonances. Using an analytical

$H_2(X^1\Sigma_g^+)$ R =10 a.u. λ_e=1.59 a.u.

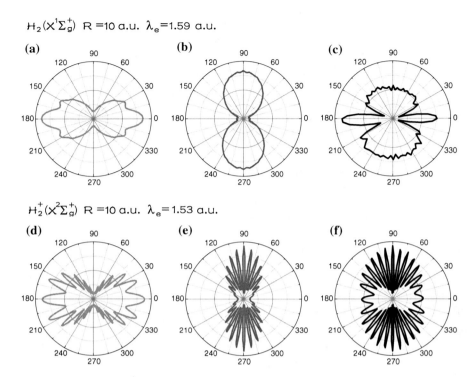

Fig. 6.9 MPADs for (**a–c**, *top row*) the H_2 $X^1\Sigma_g^+$ and (**d–f**, *bottom row*) the H_2^+ $X^2\Sigma_g^+$ ground electronic states at $\lambda = 5$ nm, $R = 10$ a.u., $\lambda_e < R$, for (**a, d**, *left column*) parallel *linear* polarization, (**b, e**, *middle column*) perpendicular *linear* polarization, and (**c, f**, *right column*) *circular* polarization ionizations

approximation to the two center Coulomb wavefunction, the asymmetric interference in HHe^{2+} has also been studied [75]. In comparison to the simplest symmetric molecular ion H$_2^+$ with a completely delocalized electron distributions, in the asymmetric molecular ion HHe^{2+}, the charge is mainly localized on the heavier He^{2+} nucleus where the electronic wave function of the ground $1s\sigma$ state is mainly concentrated. We show that in circularly polarized attosecond XUV laser pulses the HHe^{2+} MPADs display an asymmetry due to the effects of the localized electron configurations of each atom [73–75]. We have found that the angular distributions of photoelectron spectra in parallel and perpendicular (to the molecular axis) photoionizations are functions of the laser pulse wavelengths, which vary the photoelectron energy and wavelength as well.

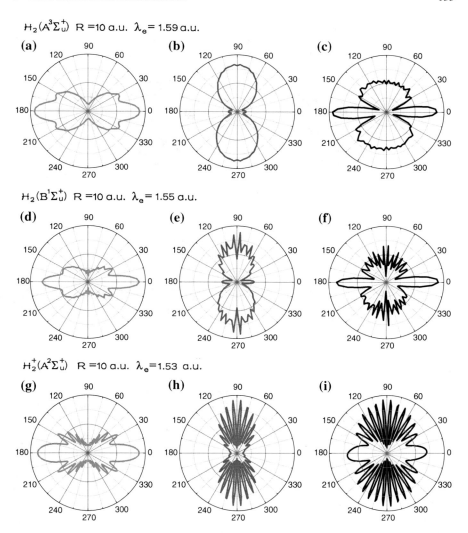

Fig. 6.10 MPADs for (**a–c**, *top row*) the H_2 $A^3\Sigma_u^+$ and (**d–f**, *middle row*) the $B^1\Sigma_u^+$ and (**g–i**, *bottom row*) the H_2^+ $A^2\Sigma_u^+$ excited electronic states at $\lambda = 5$, and $R = 10$ a.u., $\lambda_e < R$, for (**a, d, g**, *left column*) parallel *linear* polarization, (**b, e, h**, *middle column*) perpendicular *linear* polarization, and (**c, f, i**, *right column*) *circular* polarization ionizations

6.4.1 MPAD in Asymmetric Molecules

The results for the asymmetric molecular ion HHe^{2+} ground $1s\sigma$ electronic state, essentially the $He^+(1s)$ orbital, are illustrated in Fig. 6.11 at the laser wavelengths $\lambda = 15$ nm and duration $T = 300$ as at $R = 2$ a.u. and $\lambda = 20$ nm and $T = 400$ as at $R = 10$ a.u.. In Fig. 6.12 we show the momentum and angular distributions for

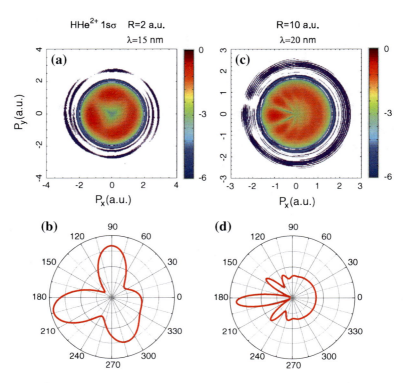

Fig. 6.11 2D momentum (p_x, p_y) distributions in *circular* polarization ionization for the HHe^{2+} ground $1s\sigma$ electronic state at internuclear distance. **a** $R = 2$ and **c** $R = 10$ a.u., ionized respectively by $\lambda = 15$ nm ($\omega = 3.04$ a.u.) and 20 nm (2.28 a.u.) attosecond XUV laser pulses with duration $T = 6$ o.c. and intensity $I_0 = 1.0 \times 10^{14}$ W/cm^2. **b** and **d** the corresponding MPADs. The corresponding wavelengths of the ionized electron at energies E_{e1} are $\lambda_{e1} = 4.59$ a.u. $\approx 2.3R$ ($R = 2$ a.u.) and $\lambda_{e1} = 5.61$ a.u. $\approx 0.56R$ ($R = 10$ a.u.)

the HHe^{2+} first excited $2p\sigma$ electronic state [76], i.e., the localized H(1s) orbital, with the same laser pulses used in Fig. 6.11 but different wavelengths and durations. This state was shown to induce resonant high order harmonics at 400 nm intense linearly polarized laser excitation [73, 74]. In Fig. 6.12a and b at $R = 2$ a.u., the pulse wavelength is $\lambda = 20$ nm and duration is $T = 400$ as whereas in Fig. 6.12c and d at $R = 10$ a.u., $\lambda = 30$ and $T = 600$ as. In the case of the photoionization of HHe^{2+}, the momentum and angular distributions show signature of large backward-forward asymmetry. In Figs. 6.11 and 6.12 at $R = 2$ a.u., the corresponding electron wavelengths are $\lambda_e = 4.59$ a.u. and 4.63 a.u. where $\lambda_e > R = 2$ a.u., thus no diffraction occurs. At $R = 10$ a.u., $\lambda_e \approx R/2$, the effect of LIED leads to a "wing" structure in momentum and angular distributions at different angles. We note that for HHe^{2+} in Figs. 6.11 and 6.12 at $R = 10$ a.u. the "wing" diffraction fingers are critically sensitive to localization of the electron, i.e., "jets" appear backward (towards H$^+$, $\theta = 180°$) for the ground $1s\sigma$ state where the electron is localized on

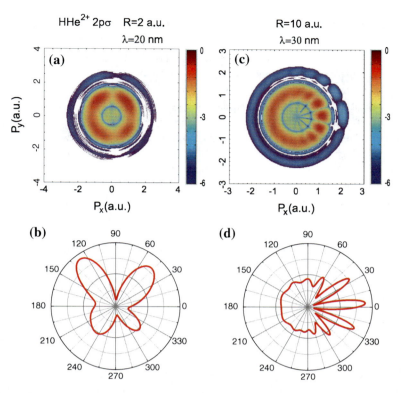

Fig. 6.12 2D momentum (p_x, p_y) distributions in *circular* polarization ionization for the HHe^{2+} first excited $2p\sigma$ electronic state at internuclear distance. **a** $R = 2$ and **c** $R = 10$ a.u., ionized respectively by $\lambda = 20$ nm ($\omega = 2.28$ a.u.) and 30 nm (1.52 a.u.) attosecond XUV laser pulses with duration $T = 6$ o.c. and intensity $I_0 = 1.0 \times 10^{14}$ W/cm^2. **b** and **d** the corresponding MPADs. The corresponding wavelengths of the ionized electron at energies E_{el} are $\lambda_{el} = 4.63$ a.u. $\approx 2.31R$ ($R = 2$ a.u.) and $\lambda_{el} = 5.24$ a.u. $\approx 0.52R$ ($R = 10$ a.u.)

He$^+$(1s) (Fig. 6.11d) and forward (towards He^{2+}, $\theta = 0°$) for the first excited $2p\sigma$ state where the electron is localized on H(1s) (Fig. 6.12d) ionization processes [75].

6.4.2 Interpretation of Asymmetric LIED

According to the LIED model in Appendix 1, we see indeed that in Figs. 6.11 and 6.12 for the internuclear distance $R = 2$ a.u., since $\lambda_e > R$, then $p_e R/2 = \pi R/\lambda_e < \pi$. The maxima of the interference terms $\cos(\mathbf{p}_e \cdot \mathbf{R}/2) = \cos(p_e R \cos \theta_{pR}/2)$ should appear at $\cos \theta_{pR} = \lambda_e/R > 1$. Then no diffraction is possible and the momentum distributions are essentially atomic. For the HHe^{2+} ground $1s\sigma$ electronic state in Fig. 6.11a and b an asymmetric distribution is due to the asymmetric initial electronic wavefunctions, $\psi \approx c_2 \psi_2 = \psi_{1s}^{He}$, whereas for the HHe^{2+} first excited $2p\sigma$ state at $R = 2$ a.u., $c_1 \psi_1 = \psi_{1s}^{H} \sim c_2 \psi_2 = \psi_{1s}^{He}$ slight asymmetric MPADs are obtained in Fig. 6.12a and b. The indication of diffraction appears in MPADs of Figs. 6.11 and 6.12 at large internuclear distance $R = 10$ a.u.. For the HHe^{2+} ground $1s\sigma$ electronic state which is mainly ψ_{1s}^{He} in Fig. 6.11, the electron is excited from a localized atomic orbital, and the distribution follows the forms $\cos(p_e R \cos\theta_{pR}/2)$. At $R = 10$ a.u. and $\lambda = 20$ nm, $p_e R \cos\theta_{pR}/2 = \pi R \cos\theta_{pR}/\lambda_e$, then $\cos(p_e R \cos\theta_{pR}/2)$ is maximum at $\cos\theta_{pR} = 0, \pm 0.56$ and ± 1. "Wings" occur near $\theta_{pR} = n\pi/2, n\pi \pm 0.31\pi$, i.e., $0°, \pm 60°, \pm 90°, \pm 120°$, and $180°$, as shown in Fig. 6.11d. For the HHe^{2+} $2p\sigma$ state at $R = 10$ a.u. with $\psi = \psi_{1s}^{H}$, maxima occur at $\theta_{pR} = n\pi/2, n\pi \pm 0.324\pi$ in Fig. 6.12d as well. For the cases of $R = 10$ a.u. for both H$_2^+$ and HHe^{2+} LIED angle nodes mainly result from the direct photoionization. We reemphasize that for the HHe^{2+} $1s\sigma$ ground electronic state, the initial electron wave packet is mainly localized on the He^{2+} ion. Due to the Coulomb effects of the H$^+$ ion, backward ($\theta = 180°$), ionization dominates [75], i.e., towards H$^+$. As shown in Fig. 6.11, the backward distributions have approximately twice the forward intensity. For the HHe^{2+} $2p\sigma$ electronic excited state at $R = 2$ a.u. where the electron density on H$^+$ and He^{2+} ions are comparable [73], near equal ionization occurs in forward and backward directions in Fig. 6.12a and b. However, at $R = 10$ a.u. the electron density on the H$^+$ ion dominates due to charge transfer, leading to a strong signal intensity of ionization along the forward direction, as illustrated in Fig. 6.12c and d.

Of note is that at $\theta_{pR} = 140°$ and $220°$ in Fig. 6.11d and $\theta_{pR} = 35°$ and $335°$ in Fig. 6.12d, clears nodes are also observed, which do not follow the LIED model. To understand the additional nodes in the MPADs in Figs. 6.11d and 6.12d, in Fig. 6.13 we show the results for the HHe^{2+} ground $1s\sigma$ and excited $2p\sigma$ electron states with respectively parallel and perpendicular (to the molecular axis) linearly polarized attosecond laser pulses with wavelengths $\lambda = 20$ nm ($\omega = 2.28$ a.u.) and $\lambda = 30$ nm ($\omega = 1.52$ a.u.), intensity $I_0 = 1.0 \times 10^{14}$ W/cm^2 and duration $T = 6$ o.c.. Comparing Figs. 6.11d and 6.13a, b we see that the diffraction nodes at angle $120°$ and $240°$ mainly result from the contributions of the perpendicular linear polarization ionization, which in fact are identical to the classical Young double-slit interference whereas those at angles at $\theta_{pR} = 140°$ and $220°$ due to the parallel linear polarizations. For the parallel linear polarization (along the molecular axis) cases, due to the strong asymmetry of the atomic distributions, the scattering effects of the ionized electron by the nuclei become more prominent and change the angular momentum of the electron, thus resulting in additional nodes in final MPADs and the electron diffraction plays a minor role in the photoionization processes. The similar processes occur for the HHe^{2+} $2p\sigma$ electronic state in Figs. 6.12d and 6.13c, d. Therefore for the asymmetric molecular ion HHe^{2+} at large internuclear distance $R = 10$ a.u., the additional angle nodes in photoelectron distributions illustrated in Figs. 6.11c, d and 6.12c, d mainly result from the atomic-like scattering effects of the ionized electron [77].

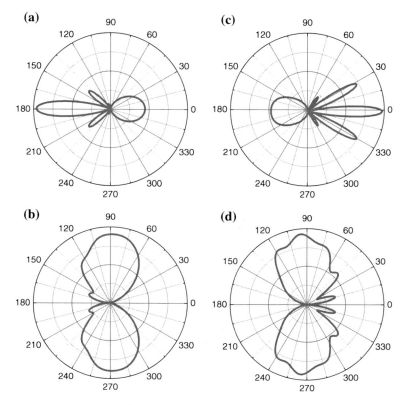

Fig. 6.13 MPADs for the HHe^{2+} (**a, b**) ground $1s\sigma$ and (**c, d**) excited $2p\sigma$ electronic states at internuclear distance $R = 10$ a.u. in intensity $I_0 = 1.0 \times 10^{14}$ W/cm^2 and duration $T = 6$ o.c. attosecond XUV laser pulses with wavelengths (**a, b**) $\lambda = 20$ nm ($\omega = 2.28$ a.u.) and (**c, d**) $\lambda = 30$ nm ($\omega = 1.52$ a.u.), for (**a, c**) parallel and (**b, c**) perpendicular *linear* polarizations, corresponding to Figs. 6.11c, d and 6.12c, d

6.5 Dependence of MATI Spectra on Laser Frequency

6.5.1 MPAD in MATI with XUV Pulses

In Fig. 6.14 we show the MPADs $\mathcal{J}^{E_e}(\theta)$ at particular MATI electron energies $E_{e1} = \omega - I_p$ and $E_{e2} = 2\omega - I_p$ for the asymmetric HHe^{2+} ground $1s\sigma$ electronic states in circularly polarized XUV laser pulses with different photon energies ω. The photoelectron energies E_{e1} and E_{e2} are the result of one and two photon ionizations. For comparison, in Fig. 6.15 we also illustrate the results for the symmetric H$_2^+$ ground $1s\sigma_g$ state. From Figs. 6.14 and 6.15 we see that the intensities of the parallel ($\theta = 0°$, and $180°$) and perpendicular ($\theta = 90°$ and $270°$) (to the molecular axis) MPADs are sensitive to the laser pulse wavelength λ and photoelectron energy E_{en}.

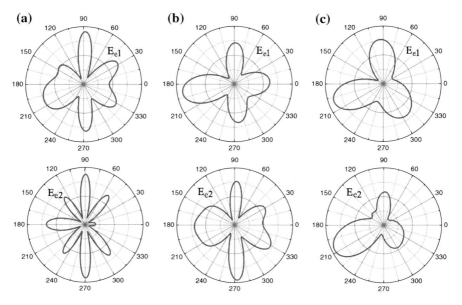

Fig. 6.14 MPADs for circular polarization ionization at specific photoelectron kinetic energies $E_{e1} = \omega - I_p$ (*left column*) and $E_{e2} = 2\omega - I_p$ (*right column*) for the HHe^{2+} ground $1s\sigma$ electronic state at internuclear distance $R = 2$ a.u. in the intensity $I_0 = 1.0 \times 10^{14}$ W/cm^2 and duration $T = 6$ o.c. attosecond XUV laser pulses with wavelengths. **a** $\lambda = 5$ nm ($\omega = 9.1$ a.u.), **b** 10 nm (4.56 a.u.), and **c** 20 nm (2.28 a.u.)

For H$_2^+$ in Fig. 6.15, at energy E_{e1} (left column) the parallel distributions decrease in intensity with increase of wavelengths λ of laser pulses. For $\lambda = 10$ nm laser pulses, the parallel and perpendicular distributions are comparable, whereas at $\lambda = 30$ nm perpendicular distributions dominate and the parallel ones nearly disappear. Thus at the shorter wavelength $\lambda = 10$ nm, the electron kinetic energies E_e are higher and less influenced by Coulomb potentials. At energy E_{e2} where a second photon is absorbed (right column) MPADs show clear nodes in both parallel and perpendicular directions due to shorter electron wavelengths $\lambda_e \simeq R$. At $\lambda = 10$ nm due to effects of the electron diffraction six nodes are obtained. Comparing results for HHe^{2+} in Fig. 6.14 and H$_2^+$ in Fig. 6.15 and shows different angular distributions. Comparable intensities in the parallel and perpendicular MPADs are obtained in Fig. 6.14, nearly independent on the pulse wavelength λ. The asymmetry in angular distributions mainly results from the initial electron configurations, as derived in (6.15). At $\lambda = 5$ nm and energy E_{e2} we get nearly symmetric diffraction "wings" in the perpendicular distributions ($\theta = 90°$ and $270°$).

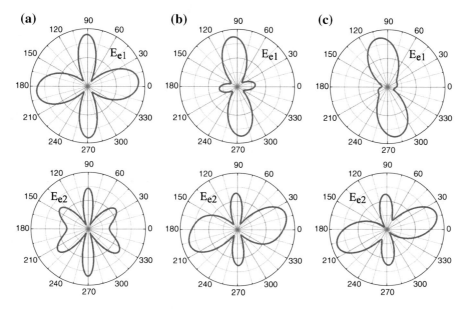

Fig. 6.15 MPADs for circular polarization ionization at specific photoelectron kinetic energies $E_{e1} = \omega - I_p$ (*left column*) and $E_{e2} = 2\omega - I_p$ (*right column*) for the H$_2^+$ ground $1s\sigma_g$ electronic state at internuclear distance $R = 2$ a.u. in the intensity $I_0 = 1.0 \times 10^{14}$ W/cm^2 and duration $T = 6$ o.c. attosecond XUV laser pulses with wavelengths. **a** $\lambda = 10$ nm ($\omega = 4.56$ a.u.), **b** 20 nm (2.28 a.u.), and **c** 30 nm (1.52 a.u.)

6.5.2 Orientation Dependent Ionization Probability

In order to understand such pulse wavelength λ_e and photoelectron kinetic energy E_e dependent MPADs in Figs. 6.14 and 6.15, in Fig. 6.16 we show the ratio of parallel to perpendicular linear polarization probabilities at particular photoelectron kinetic energies E_e with varying wavelength λ of laser pulses for the H$_2^+$ ground $1s\sigma_g$ and the HHe^{2+} ground $1s\sigma$ electronic states. We define this ratio Γ as

$$\Gamma = \frac{\mathscr{J}_{\parallel}^{E_e} - \mathscr{J}_{\perp}^{E_e}}{\mathscr{J}_{\parallel}^{E_e} + \mathscr{J}_{\perp}^{E_e}}, \tag{6.12}$$

where $\mathscr{J}_{\parallel}^{E_e}$ and $\mathscr{J}_{\perp}^{E_e}$ are respectively differential yields obtained by parallel and perpendicular linearly polarized laser pulses. In Fig. 6.16a for the H$_2^+$ $1s\sigma_g$ electronic state we see the ratio as functions of both the pulse wavelength λ and the photoelectron kinetic energy E_e. An oscillatory behavior of the ratio Γ in (6.12) is observed. At $E_{e1} = \omega - I_p$ the ratio Γ gradually increases as λ decreases. As seen in Fig. 6.16a Γ is nearly -1 for the long pulse wavelengths $\lambda \geq 30$ nm with low photon energy ω and long electron wavelength $\lambda_e \gg R$, i.e., the perpendicular ionization

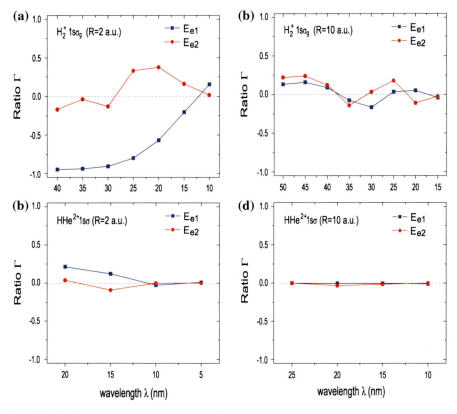

Fig. 6.16 Ratio Γ of photoionization probabilities obtained from (6.12) at specific photoelectron kinetic energies E_e for parallel and perpendicular *linear* polarization ionization by attosecond XUV laser pulses with duration $T = 6$ o.c. and intensity $I_0 = 1.0 \times 10^{14}$ W/cm^2 as a function of wavelength λ for (**a, c**) the H$_2^+$ $1s\sigma_g$ and (**b, d**) the HHe^{2+} $1s\sigma$ electronic states at internuclear distances (**a, b**) $R = 2$ a.u. and (**c, d**) $R = 10$ a.u.

distributions dominate, whereas for the cases of shorter pulse λ with higher photon energy ω inducing shorter electron wavelengthes $\lambda_e \sim R$, Γ is closed to 0, indicating approximately equal photoelectron distributions in parallel and perpendicular directions. At energy $E_{e2} = 2\omega - I_p$ where the wavelength $\lambda_e \leq 2R$ the ratio Γ slightly oscillates around 0. The parallel and perpendicular distributions are comparable in intensity. These results consistent with the MPADs in circularly polarized laser fields as illustrated in Fig. 6.15.

The laser wavelength λ and photoelectron kinetic energy E_e dependence of the ratio Γ mainly results from the effects of the electronic charge clouds associated with each scattering center in the photoionization processes, leading to orientation dependent ionization rates [78]. For the photoionization with long laser wavelengths producing long electron wavelength λ_e, the oriented ionization effects are significant. Decreasing photoelectron wavelength λ_e, the orientation effects on the

6 Laser Induced Electron Diffraction   143

ionization processes weaken due to weaker Coulomb potential effects. The molecular photoionization is atomic-like. As a result, the ratio Γ is slightly sensitive to the pulse wavelength λ and photoelectron energy E_e for the H_2^+ ground $1s\sigma_g$ electronic state as shown in Figs. 6.15 and 6.16a. Under conditions $\lambda_e < R$ the effects of the two center diffractions of the photoelectron are clearly observed, as predicted in (6.15). The photon and photoelectron energy dependence arises from the Coulomb scattering of the electron wave packets and therefore changes of angular momentum of the electron, leading to different MPADs [77].

In Fig. 6.16b we show the results for the HHe^{2+} ground $1s\sigma$ electronic state at $R = 2$ a.u.. It clearly shows that the degree of the ratio Γ is insensitive to the pulse wavelength λ, and is nearly constant at $\Gamma \approx 0$ for both photoelectron kinetic energies E_{e1} and E_{e2}. The oscillatory dependence with λ in Fig. 6.16a for equilibrium H_2^+ is nearly absent. For the HHe^{2+} $1s\sigma$ electronic state, the initial electronic density is mainly localized on the He^{2+} ion, leading to an atomic electron distribution. The orientation dependent photoionization from the geometrical effects is therefore absent. Thus the ratio Γ of the parallel to perpendicular photoionization probabilities in Fig. 6.16b is not modified by a change in the wavelength λ and photoelectron kinetic energy E_e (λ_e) for HHe^{2+}.

Figure 6.16c and d illustrate the ratio Γ for the H_2^+ $1s\sigma_g$ and the HHe^{2+} $1s\sigma$ states at large internuclear distance $R = 10$ a.u.. The orientation effects on the pulse wavelength λ dependence of photoionization nearly disappear. For the H_2^+ $1s\sigma_g$ state, a slight oscillatory structure around $\Gamma = 0$ is observed with pulse wavelength λ at both photoelectron kinetic energies E_{e1} and E_{e2}. This differs from the equilibrium results in Fig. 6.16a at $R = 2$ a.u.. For the extended molecular ion with large $R = 10$ a.u., the initial electronic state involves an electron delocalized equally over the two centers. Then the orientational effects on the photoionizations is negligible. In comparison, in Fig. 6.16d photoionizations for the HHe^{2+} $1s\sigma$ state show nearly equal ionization probabilities in parallel and perpendicular polarizations ($\Gamma = 0$). These are still independent of the laser pulse wavelengths λ and photoelectron energies E_e and are identical to $R = 2$ a.u. photoionization confirming complete atomic localization of the electron on the He^{2+} ion.

6.6 Conclusions

In this chapter we summarize manifestations of LIED in MAP with circularly polarized XUV laser pulses from numerical solutions of the corresponding TDSEs for static nuclei. Simulations are performed on oriented symmetric molecules H_2 and H_2^+, and asymmetric HHe^{2+}. Under proper conditions, MPADs calculated by analyzing the associated electron current densities exhibit diffraction patterns.

We show that MPADs display dominant single-electron diffraction effects whenever the ionized electron wavelength λ_e is less than the internuclear distance R with electron "jets" emitted at particular angles. In general, for the condition of $\lambda > R$ where no diffraction occurs, circular polarization MPADs are sums of linear parallel and perpendicular (to the molecular R axis) angular distributions [60]. The MPADs are analyzed using an exactly solvable photoionization model by a delta function pulse in LIED [6]. The MPAD amplitudes obtained by Fourier transform of the electron wave functions show that the MPAD is regulated by the double-slit interference term $\cos(\mathbf{p} \cdot \mathbf{R})/2$ or $\sin(\mathbf{p} \cdot \mathbf{R})/2$ for g or u electronic states multiplied by an atomic angular term. At larger internuclear distance $R = 10$ a.u due to the highly delocalized nature of an electron over the two separated protons for a single electron in H_2^+, the angular and momentum distributions for $X^2 \Sigma_g^+$ and $A^2 \Sigma_u^+$ electronic states show strong similarities. The H_2 ground $X^1 \Sigma_g^+$ and the excited $A^3 \Sigma_u^+$ electronic states show very similar distributions for linear parallel and perpendicular polarization due to same Heitler-London atomic orbital configurations at large dissociation distance. Circular polarization ionizations are not simple superposition of linear parallel and perpendicular polarization results. This suggests the effects of electron correlation and entanglement in different states. Due to the different atomic electron localization of electronic states in HHe^{2+}, asymmetric angular distributions occur in this asymmetric molecule, as a result of the Coulomb potential from the other atom.

It is also found that the relative intensity of the parallel and perpendicular (to molecular R axis) MPADs of the photoelectron spectra is a function of the laser wavelength λ, photoelectron kinetic energies E_e and molecular orbital symmetry. For the symmetric equilibrium H_2^+ ground electronic state where the electron distribution is delocalized, photoelectron angular distributions at the low photoelectron energy are critically sensitive to the pulse wavelength λ since the ionization rates depend on molecular geometry, whereas at high energy and short wavelength $\lambda_e < R$ orientation effects on ionization nearly disappear and effects of LIED dominate. For the asymmetric HHe^{2+} $1s\sigma$ electron state where the electron distribution is highly localized on the He^{2+} ion, these effects are nearly absent.

Acknowledgments The authors thank RQCHP and Compute Canada for access to massively parallel computer clusters and CIPI (Canadian Institute for Photonic Innovations) for financial support of this research in its ultrafast science program. A. D. Bandrauk is also indebted to a Canada Research Chair for pursuing attosecond research.

Appendix 1

For a single electron diatomic molecule at internuclear distance R, the initial electronic state can be written in a general form with respect to the center of mass of two nuclei

$$\psi^{\pm}(\mathbf{R}, \mathbf{r}) = c_1 \psi_1(\mathbf{r} + \mathbf{R}_1,) \pm c_2 \psi_2(\mathbf{r} - \mathbf{R}_2,), \qquad (6.13)$$

where $R_1 = m_1 R/(m_1 + m_2)$ and $R_2 = m_2 R/(m_1 + m_2)$ are the positions of the two nuclei with masses m_1 and m_2 and $\mathbf{R} = \mathbf{R}_1 - \mathbf{R}_2$. In linearly polarized laser pulses, the ionization distribution is mainly along the polarization of the driving laser pulse. After excitation by a $\delta-$pulse with linear $\mathbf{e}$ polarization which defined as,

6 Laser Induced Electron Diffraction

$$\mathbf{E}(t) = \mathbf{F}\delta(t = 0),\qquad(6.14)$$

where $\mathbf{F}$ is the pulse amplitude, the corresponding MPADs are obtained by a Fourier transform of the electron wave function $\psi^{\pm}(\mathbf{R}, \mathbf{r}, t = 0^{+})$,

$$
\begin{aligned}
f^{\pm}(\mathbf{p}) &\propto (\mathbf{e} \cdot \mathbf{n})^2 |\mathscr{F}[\psi^{\pm}(\mathbf{R}, \mathbf{r}, t = 0^{+})]|^2 \\
&\propto (\mathbf{e} \cdot \mathbf{n})^2 |\int \exp(-i\mathbf{p} \cdot \mathbf{r}) \exp(-i\mathbf{F} \cdot \mathbf{r}) \psi^{\pm}(\mathbf{R}, \mathbf{r}) d\mathbf{r}|^2 \\
&= (\mathbf{e} \cdot \mathbf{n})^2 |c_1 \psi_1(|\mathbf{F} + \mathbf{p}|)\{\cos[(\mathbf{F} + \mathbf{p}) \cdot \mathbf{R}_1] + i \sin[(\mathbf{F} + \mathbf{p}) \cdot \mathbf{R}_1]\} \\
&\pm c_2 \psi_2(|\mathbf{F} + \mathbf{p}|)\{\cos[(\mathbf{F} + \mathbf{p}) \cdot \mathbf{R}_2] - i \sin[(\mathbf{F} + \mathbf{p}) \cdot \mathbf{R}_2]\}|^2 \\
&= (\mathbf{e} \cdot \mathbf{n})^2 \{a + b \cos^2[(\mathbf{F} + \mathbf{p}) \cdot \mathbf{R}/2]\},\qquad(6.15)
\end{aligned}
$$

where $\mathbf{p}$ is photoelectron momentum and the unit vector $\mathbf{n} = \mathbf{p}/p$, $a = [c_1 \psi_1(|\mathbf{F} + \mathbf{p}|) - c_2 \psi_2(|\mathbf{F} + \mathbf{p}|)]^2$ and $b = 4c_1 c_2 \psi_1(|\mathbf{F} + \mathbf{p}|)\psi_2(|\mathbf{F} + \mathbf{p}|)$. The MPADs are the interaction of the electron with the photon $(\mathbf{e} \cdot \mathbf{n})$ and the two center nature of the initial state $(a + b \cos^2[(\mathbf{F} + \mathbf{p}) \cdot \mathbf{R}/2])$. Of note is that in the classical diffraction model the interference of the scattering electron wave packet by the two nuclear centers in the final state is omitted as this interference dose not significantly influence MPADs [30]. For symmetric molecular electronic state, the MPADs in (6.15) reduce to

$$f_g(\mathbf{p}) = 2(\mathbf{e} \cdot \mathbf{n})^2 \cos^2[(\mathbf{F} + \mathbf{p}) \cdot \mathbf{R}/2]\psi(|\mathbf{F} + \mathbf{p}|)^2,\qquad(6.16)$$

and for the antisymmetric electronic state,

$$f_u(\mathbf{p}) = 2(\mathbf{e} \cdot \mathbf{n})^2 \sin^2[(\mathbf{F} + \mathbf{p}) \cdot \mathbf{R}/2]\psi(|\mathbf{F} + \mathbf{p}|)^2.\qquad(6.17)$$

For asymmetric MPADs are linear combinations of the appropriate atomic amplitudes with coefficients c_1 and c_2 on each ions. Under conditions with negligible ponderomotive energies $U_p = E^2/4\omega^2 \ll I_p$, we can use the results in 6.15, 6.16 and 6.17 with $\mathbf{F} = 0$. Moreover effects of the laser induced Stark shift can also be ignored.

References

1. T. Brabec, F. Krausz, Rev. Mod. Phys. **72**, 545 (2000)
2. F. Krausz, M. Ivanov, Rev. Mod. Phys. **81**, 163 (2009)
3. Z. Chang, P. Corkum, J. Opt. Soc. Am. B **27**, B9 (2010)
4. S. Chelkowski, P.B. Corkum, A.D. Bandrauk, Phys. Rev. Lett. **82**, 3416 (1999)
5. F. Légaré, K.F. Lee, A.D. Bandrauk, D.M. Villeneuve, P.B. Corkum, J. Phys. B: At. Mol. Opt. Phys. **39**, S503 (2006)
6. T. Zuo, A.D. Bandrauk, P.B. Corkum, Chem. Phys. Lett. **259**, 313 (1996)
7. M. Meckel, D. Comtois, D. Zeidler, A. Staudte, D. Pavicic, H.C. Bandulet, H. Pépin, J.C. Kieffer, R. Dörner, D.M. Villeneuve, P.B. Corkum, Science **320**, 1478 (2008)

8. M. Lein, J.P. Marangos, P.L. Knight, Phys. Rev. A **66**, 051404 (2002)
9. S.X. Hu, L.A. Collins, Phys. Rev. Lett. **94**, 073004 (2005)
10. M. Peters, T.T. Nguyen-Dang, C. Cornaggia, S. Saugout, E. Charron, A. Keller, O. Atabek, Phys. Rev. A **83**, 051403(R) (2011)
11. M. Peters, T.T. Nguyen-Dang, E. Charron, A. Keller, O. Atabek, Phys. Rev. A **83**, 053417 (2011)
12. M. Spanner, O. Smirnova, P.B. Corkum, M.Y. Ivanov, J. Phys. B: At. Mol. Opt. Phys. **37**, L243 (2004)
13. K. Zhao, Q. Zhang, M. Chini, Y. Wu, X. Wang, Z. Chang, Opt. Lett. **37**, 3891 (2012)
14. A.D. Bandrauk, S. Chelkowski, D.J. Diestler, J. Manz, K.J. Yuan, Int. J. Mass Spectrom. **277**, 189 (2008)
15. M. Vrakking, Nature **460**, 960 (2009)
16. M.J.J. Vrakking, Physics **2**, 72 (2009)
17. H.D. Cohen, U. Fano, Phys. Rev. **150**, 30 (1966)
18. I.G. Kaplan, A.P. Markin, Sov. Phys. Dokl. **14**, 36 (1969)
19. S.N. Yurchenko, S. Patchkovskii, I.V. Litvinyuk, P.B. Corkum, G.L. Yudin, Phys. Rev. Lett. **93**, 223003 (2004)
20. P.B. Corkum, Phys. Rev. Lett. **71**, 1994 (1993)
21. A.D. Bandrauk, S. Chelkowski, S. Goudreau, J. Mod. Opt. **52**, 411 (2005)
22. A.D. Bandrauk, S. Chelkowski, H. Yu, E. Constant, Phys. Rev. A **56**, R2537 (1997)
23. A.D. Bandrauk, S. Barmaki, G.L. Kamta, Phys. Rev. Lett. **98**, 013001 (2007)
24. M. Okunish, R. Itaya, K. Shimada, G. Prümper, K. Ueda, M. Busuladžić, A. Gazibegovići-Busuladžić, D.B. Milošević, W. Becker, Phys. Rev. Lett. **103**, 043001 (2009)
25. K.J. Yuan, A.D. Bandrauk, Phys. Rev. A **84**, 013426 (2011)
26. K.J. Yuan, A.D. Bandrauk, Phys. Rev. A **84**, 023410 (2011)
27. K.J. Yuan, A.D. Bandrauk, Phys. Rev. A **81**, 063412 (2010)
28. K.J. Yuan, A.D. Bandrauk, Phys. Rev. A **83**, 063422 (2011)
29. D.A. Horner, S. Miyabe, T.N. Rescigno, C.W. McCurdy, F. Morales, F. Martín, Phys. Rev. Lett. **101**, 183002 (2008)
30. A.S. Baltenkov, U. Becker, S.T. Manson, A. Msezane, J. Phys. B: At. Mol. Opt. Phys. **45**, 035202 (2012)
31. S.E. Cantona, E. Plésiatb, J.D. Bozekc, B.S. Ruded, P. Declevae, F. Martín, Proc. Natl. Acad. Sci. USA **108**, 7302 (2011)
32. U. Becker, Science **474**, 586 (2011)
33. I. Barth, J. Manz, Angew. Chem. Int. Ed. **45**, 2962 (2006)
34. I. Barth, C. Bressler, S. Koseki, J. Manz, Chem. Asian J. **7**, 1261 (2012)
35. C.P.J. Martiny, L.B. Madsen, Phys. Rev. Lett **97**, 093001 (2006)
36. C.P.J. Martiny, M. Abu-Samha, L.B. Madsen, Phys. Rev. A **81**, 063418 (2010)
37. D. Akoury, K. Kreidi, T. Jahnke, Th Weber, A. Staudte, M. Schöffler, N. Neumann, J. Titze, LPhH Schmidt, A. Czasch, O. Jagutzki, R.A.C. Fraga, R.E. Grisenti, R.D. Muiño, N.A. Cherepkov, S.K. Semenov, P. Ranitovic, C.L. Cocke, T. Osipov, H. Adaniya, J.C. Thompson, M.H. Prior, A. Belkacem, A.L. Landers, H. Schmidt-Böcking, R. Dörner, Science **318**, 949 (2007)
38. H.K. Kreidi, D. Akoury, T. Jahnke1, Th. Weber, A. Staudte, M. Schöffler, N. Neumann, J. Titze, L.Ph.H. Schmidt, A. Czasch, O. Jagutzki1, R.A.C. Fraga, R.E. Grisenti, M. Smolarski, P. Ranitovic, C.L. Cocke, T. Osipov, H. Adaniya, J.C. Thompson, M.H. Prior, A. Belkacem, A.L. Landers, H. Schmidt-Böcking, R.Dörner, Phys. Rev. Lett. **100**, 133005 (2008)
39. H. Akagi, T. Otobe, A. Staudte, A. Shiner, F. Turner, R. Dörner, D.M. Villeneuve, P.B. Corkum, Science **325**, 1364 (2010)
40. A. Staudte, S. Patchkovskii, D. Pavičć, H. Akagi, O. Smirnova, D. Zeidler, M. Meckel, D.M. Villeneuve, R. Dörner, M.Y. Ivanov, P.B. Corkum, Phys. Rev. Lett. **102**, 033004 (2009)
41. A. Fleischer, H.J. Wörner, L. Arissian, L.R. Liu, M. Meckel, A. Rippert, R. Dörner, D.M. Villeneuve, P.B. Corkum, A. Staudte, Phys. Rev. Lett. **107**, 113003 (2011)
42. L. Holmegaard, J.L. Hansen, L. Kalhøj, S.L. Kragh, H. Stapelfeldt, F. Filsinger, J.Küpper, G. Meijer, D. Dimitrovski, M. Abu-samha, C.P.J. Martiny, L.B. Madsen, Nature Phys. **6**, 428 (2010)

6 Laser Induced Electron Diffraction

43. X. Zhu, Q. Zhang, W. Hong, P. Lu, Z. Xu, Opt. Express **19**, 13722 (2011)
44. X. Ren, J. Zhang, Y. Wu, Z. Xu, Phys. Lett. A **376**, 1889 (2012)
45. G.M. Genkin, YuN Nozdrin, A.V. Okomel'kov, I.D. Tokman, Phys. Rev. B **86**, 024405 (2012)
46. C.H.R. Ooi, W. Ho, A.D. Bandrauk, Phys. Rev. A **86**, 023410 (2012)
47. I. Barth, O. Smirnova, Phys. Rev. A **84**, 063415 (2011)
48. K.J. Yuan, A.D. Bandrauk, J. Phys. B: At. Mol. Opt. Phys. **45**, 074001 (2012)
49. K.J. Yuan, A.D. Bandrauk, Phys. Rev. Lett. **110**, 023003 (2013)
50. F. Morales, I. Barth, V. Serbinenko, S. Patchkovskii, O. Smirnova, J. Mod. Opt. **59**, 1303 (2012)
51. I. Barth, J. Manz, Y. Shigeta, K. Yagi, J. Am. Chem. Soc. **128**, 7043 (2006)
52. I. Barth, J. Manz, Phys. Rev. A **75**, 012510 (2007)
53. F. Mauger, C. Chandre, T. Uzer, Phys. Rev. Lett. **104**, 043005 (2010)
54. F. Mauger, C. Chandre, T. Uzer, Phys. Rev. Lett. **105**, 083002 (2010)
55. K.J. Yuan, A.D. Bandrauk, Phys. Rev. A **88**, 013417 (2013)
56. A.D. Bandrauk, S. Chelkowski, Phys. Rev. Lett. **87**, 273004 (2001)
57. S. Chelkowski, A.D. Bandrauk, Appl. Phys. B **74**, S113 (2002)
58. A.D. Bandrauk, H.Z. Lu, in High-Dimensional Partial Differential Equations, in *Science and Engineering*, vol. 41, ed. by A.D. Bandrauk, M. Delfour, C. LeBris, C.R.M. Lecture Series, (American Mathematical Society, Philadelphia, 2007), pp. 1–15
59. A.D. Bandrauk, H. Shen, J. Chem. Phys. **99**, 1185 (1993)
60. K.J. Yuan, H.Z. Lu, A.D. Bandrauk, Phys. Rev. A **80**, 061403(R) (2009)
61. K.J. Yuan, H.Z. Lu, A.D. Bandrauk, Phys. Rev. A **83**, 043418 (2011)
62. G. Jolicard, O. Atabek, Phys. Rev. A **46**, 5845 (1992)
63. B. Feuerstein, U. Thumm, J. Phys. B: At. Mol. Opt. Phys. **36**, 707 (2003)
64. C.F. Matta, R.J. Boyd, *The Quantum Theory of Atoms in Molecules* (Wiley-VCH, New York, 2009)
65. G. Burkard, D. Loss, D.P. DiVincenzo, Phys. Rev. B **59**, 2070 (1999)
66. G. Corongiu, E. Clementi, J. Phys. Chem. A **113**, 14791 (2009)
67. K. Harumiya, H. Kono, Y. Fujimura, I. Kawata, A.D. Bandrauk, Phys. Rev. A **66**, 043403 (2002)
68. S. Saugout, C. Cornaggia, A. Suzor-Weiner, E. Charron, Phys. Rev. Lett **98**, 253003 (2007)
69. J. Fernández, F.L. Yip, T.N. Rescigno, C.W. McCurdy, F. Martín, Phys. Rev. A **79**, 043409 (2009)
70. J. Fernández, F. Martín, New J Phys. **11**, 043020 (2009)
71. M. Walter, J. Briggs, J. Phys. B: At. Mol. Opt. Phys. **32**, 2487 (1999)
72. L. Nagy, S. Borbély, K. Póra, Phys. Lett. A **327**, 481 (2004)
73. X.B. Bian, A.D. Bandrauk, Phys. Rev. Lett. **105**, 093903 (2010)
74. X.B. Bian, A.D. Bandrauk, Phys. Rev. A **83**, 023414 (2011)
75. G.L. Yudin, S. Patchkovskii, A.D. Bandrauk, J. Phys. B: At. Mol. Opt. Phys. **39**, 1537 (2006)
76. T.G. Winter, M.D. Duncan, N.F. Lane, J. Phys. B: At. Mol. Opt. Phys. **2**, 285 (1977)
77. S. Bauch, M. Bonitz, Phys. Rev. A **78**, 043403 (2008)
78. S. Selstø M. Førre, J.P. Hansen, L.B. Madsen, Phys. Rev. Lett. **95** 093002 (2005)

Chapter 7
Coherent Electron Wave Packet, CEWP, Interference in Attosecond Photoionization with Ultrashort Circularly Polarized XUV Laser Pulses

Kai-Jun Yuan and André D. Bandrauk

Abstract Effects of coherent electron interference in attosecond photoionization with intense ultrashort circularly polarized extreme ultraviolet (XUV) laser pulses are studied. Simulations are performed on oriented one electron molecular ions H_2^+ and H_3^{2+} by numerically solving appropriate time-dependent Schrödinger equations (TDSEs). It is found that due to interference of coherent continuum scattering electron wave packets, momentum *stripes* with intervals $\Delta p^s = 2\pi/R$ are observed in angular distributions. The momentum stripes are independent of the laser polarization and wavelength, and these are always perpendicular to the molecular internuclear axis. Ionization with two color circularly polarized XUV laser pulses produces an asymmetry in angular distributions resulting from interference of coherent electron wave packets (CEWPs) from multiple pathway ionization. The multiple ionization pathway interference is shown to be sensitive to the pulse relative carrier envelope phase (CEP) and photoelectron kinetic energies. We describe these phenomena by a perturbation multi-photon ionization model, thus suggesting imaging and controlling methods for electrons in molecules on attosecond time scale.

K.-J. Yuan (✉)
Laboratoire de Chimie Théorique, Faculté des Sciences, Université de Sherbrooke,
Sherbrooke, QC J1K 2R1, Canada
e-mail: kaijun.yuan@usherbrooke.ca

A. D. Bandrauk
Canada Research Chair/Computational Chemistry & Molecular Photonics
Laboratoire de Chimie Théorique, Faculté des Sciences, Université de Sherbrooke,
Sherbrooke, QC J1K 2R1, Canada
e-mail: andre.bandrauk@USherbrooke.ca

K. Yamanouchi et al. (eds.), *Progress in Ultrafast Intense Laser Science XI*,
Springer Series in Chemical Physics 109, DOI: 10.1007/978-3-319-06731-5_7,
© Springer International Publishing Switzerland 2015

7.1 Introduction

The real time imaging of chemical reactions has attracted considerable attention by using pump-prob techniques [1]. An ultrashort pump laser pulse initiates a transition in a molecule and its time evolution is subsequently monitored after a variable time delay by a probe pulse. Time-resolved photoelectron spectroscopy is developing into a most useful tool for probing and controlling ultrafast dynamics in molecules especially in the femtosecond (1 fs $= 10^{-15}$ s) time regime associated with nuclear motion [2, 3]. Thus time-resolved imaging of nuclear dynamics during a chemical reaction has been studied recently in photodissociation [4]. Time-resolved photoelectron spectroscopy at a conical intersection [5, 6] and two-dimensional (2D) femtosecond electronic spectroscopy have been shown to remove inhomogeneous broadening in nanoparticle electron dynamics [7]. Femtosecond dynamics in general involve electron-nuclear coupled dynamics requiring extensive vibronic excited state wave packet dynamics [8].

Modern ultrafast ultrashort (few cycle) laser pulse technology [9, 10] allows to study pure electron dynamics on the electrons natural time scale, the attosecond (1 as $= 10^{-18}$ s). Attosecond pulses have now achieved a new ultrashort pulse duration of 67 as [11], thus offering a new tool to watch pure electronic quantum effects without interference from nuclear motion [12], and allowing in particular through attosecond imaging the creation of electron movies [13]. Attosecond pulses allow therefore to separate electronic and nuclear effects due to the different time scales of attosecond for electron motion and femtosecond for nuclear motion. Using extreme ultraviolet (XUV) light from high-order harmonics (HHG) [14], laser plasmas [15], and free electron lasers (FEL) [16], coherent electron wave packets (CEWPs) can be created inside atoms and molecules on the attosecond time scale and sub-nanometer dimension [17–20]. Thus electron dynamics in CEWPs can be monitored by measuring photoelectron angular and energy distributions. Many studies have been reported focusing on revealing effects of atomic and molecular orbital configurations [21–23], multi-pathway quantum interference [24–28] and electron correlations [29–31] by XUV light or combined with ultrashort intense infrared (IR) laser pulses. An alternative efficient tool has been recently proposed to observe attosecond electron motion by measuring the generated radiation, molecular high-order harmonic generation (MHOHG), as a function of the pump-probe delay in a dissociating molecule [32–34], see also a recent review [35]. XUV-pump-XUV-probe experiments have been adopted in tracing atomic coherences [36] and molecular nuclear wave packet dynamics [37]. To date most experiments have been performed with linear polarization.

Recently, *circularly* polarized attosecond laser pulses have been attracted considerable attention in investigation of electron dynamics. Circular laser polarization has been emphasized earlier to be important in strong field ionization processes [38]. It has been shown that with circularly polarized laser pulses intense electron ring currents and magnetic fields are generated in cyclic molecules [39, 40] and recollision of electrons occurs in double ionization [41, 42]. Most recently, it is found that in intense circularly polarized laser fields, molecular photoelectron angular distributions

(MPADs) exhibit "tilted" or rotated angles with respect to the polarization (molecular symmetry) axes. In tunnelling ionizations of H_2^+ with IR 800 nm circularly polarized laser pulses the rotation angle of MPADs with respect to simple quasi-static model predictions [43] is attributed to a complex electronic motion inside molecules at critical distances R_c [44, 45], where enhanced ionization (EI) has previously been predicted to occur [46–48]. The unexpected tilt angle is also shown to be sensitive to pulse wavelength λ and molecular internuclear distance R [49–51]. With circularly polarized XUV laser pulses, MPADs of H_2 have also been shown to exhibit slight rotations with respect to the polarization axes due to the nonspherical Coulomb potential and the helicity of the light [52, 53]. In multi-photon XUV ionization processes the rotation angle of MPADs is dependent on the symmetry of the initial electronic state [54] and the photoelectron energy [55]. Such rotation of MPADs is also observed in two color circularly polarized XUV laser fields [56]. We also note that optimal control of spin polarization has been achieved using a circularly polarized IR pulse in solid monolayers of MoS_2 [57].

In this review chapter, we present attosecond photoionization with circularly polarized XUV laser pulses by numerically solving the time dependent Schrödinger equation (TDSE) for single electron molecular systems. Interference of scattering CEWPs are observed in MPADs of linear H_2^+ and triangular H_3^{2+} with momentum stripes which are always perpendicular to the molecular internuclear axis. Due to multiple nuclear centers in H_3^{2+} MPADs exhibit a set of alternating bright and dark interference fringes with intensity spots. Moreover, in two color circularly polarized XUV laser fields, interference of CEWPs created from different ionization pathways leads to an asymmetry in MPADs. Nuclear motion is not discernible (measurable) due to the large bandwidth and duration shorter than the femtosecond time scale [58, 59]. We focus therefore on attosecond photoionization with static nuclei and show effects of electron interference and diffraction in MPADs. We have proposed previously methods of generating circularly polarized attosecond pulses from circularly polarized MHOHG [60–62] using the nonspherical symmetry of molecules as opposed to spherical symmetry in atoms which prohibits circularly polarized harmonics, thus making it possible to observe and control the electron interfere in circular polarization attosecond photoionization.

The chapter is organized as follows: We first present the numerical method for simulations of MPADs in Sect. 7.2. Section 7.3 displays interference effects of scattering electron wave packets with momentum stripes for both H_2^+ and H_3^{2+}. For multiple pathway ionization processes by two color XUV laser pulses interference effects resulting in asymmetry in MPADs are shown in Sect. 7.4. Finally, we summarize our finding in Sect. 7.5. Throughout this chapter, atomic units (a.u.) $e = \hbar = m_e = 1$ are used.

7.2 Numerical Methods

The TDSE for a fixed nuclei oriented single electron molecular ion is appropriately written with respect to the center of mass of protons as,

$$i\frac{\partial}{\partial t}\psi(\mathbf{r},t) = \left[-\frac{1}{2}\nabla_{\mathbf{r}}^2 + V_{en}(\mathbf{r}) + V_L(\mathbf{r},t)\right]\psi(\mathbf{r},t), \qquad (7.1)$$

$\nabla_{\mathbf{r}}^2$ is the Laplacian. The cylindrical coordinates (ρ,θ,z) are adopted to describe the TDSE. $V_{en}(\mathbf{r})$ is the electron-proton attraction Coulomb potentials. The nuclear coordinates $(x,y,z) = (\rho\cos\theta, \rho\sin\theta, z) = (N_{nx}, N_{ny}, 0)$ for H_2^+ are fixed at $(\pm R/2\cos\vartheta, \pm R/2\sin\vartheta, 0)$, where ϑ is the alignment angle between the molecular axis and the x−axis. For symmetric H_3^{2+} the proton positions are in equilateral geometry [63], i.e., the coordinate relations $N_{1x} = 0$, $N_{2x} = -R/2$, $N_{3x} = +R/2$, $N_{1y} = \sqrt{3}R/3$, $N_{2y} = -\sqrt{3}R/6$, $N_{3y} = -\sqrt{3}R/6$, R being the distance between two protons. The TDSE is described using cylindrical coordinates (ρ,θ,z) with $(x = \rho\cos\theta, y = \rho\sin\theta)$, where the radial and angular variables are separated from z perpendicular to the molecular (x,y) and laser polarization plane. For 3D models the corresponding molecular ionization potentials are $I_p(H_2^+) = 1.1$ a.u. at equilibrium R_e and $I_p(H_3^{2+}) = 1.08$ a.u. at $R = 4$ a.u.

The time-dependent laser-electron radiative coupling potential is for circularly polarization

$$V_L = E_0 f(t)(\rho\cos\theta\cos\omega t + \rho\sin\theta\sin\omega t) \qquad (7.2)$$

or for linear polarization

$$V_L = E_0 f(t)\rho\cos\theta\cos\omega t \qquad (7.3)$$

in length gauge for an $\mathbf{E}(t)$ field, propagating along the z axis perpendicular to the molecular (x,y) and laser polarization plane. A temporal slowly varying envelope $f(t) = \sin^2(\pi t/T)$ with duration $T = n_l\tau$ (1 optical cycle, o.c., $\tau = 2\pi/\omega$) and maximum amplitude E_0 for intensity $I_0 = \frac{1}{2}c\varepsilon_0 E_0^2$ is adopted, so that the total pulse area $A(t) = -\frac{1}{c}\int_{t_i}^{t_f} E(t)dt$, where t_i is any time before the start of the pulse envelope and t_f is any time after the turn off of the pulse envelope, and $A(t_f) - A(t_i) = 0$ which is constrained in accordance with Maxwell's equation [64]. The TDSE is solved numerically by a second-order split-operator method in the time step δt combined with a fifth order finite difference method and Fourier transform technique in the spatial steps $\delta\rho$, $\delta\theta$, and δz [65, 66]. The time step is taken to be $\delta t = 0.01$a.u. $= 0.24$ as, the spatial discretization is $\delta\rho = \delta z = 0.25$ a.u. for radial grid ranges $0 \le \rho \le 128$ a.u., $|z| \le 64$ a.u., and the angle grid size $\delta\theta = 0.025$ radian. To prevent unphysical effects due to the reflection of the wave packet from the boundary, we multiply $\psi(\rho,\theta,z,t)$ by a "mask function" or absorber in the radial coordinate with the form

7 Coherent Electron Wave Packet
153

$$\mathcal{G}(t) = \begin{cases} 1, & \rho < \rho_a \\ \cos^{1/8}[\pi(\rho - \rho_a)/2\rho_{abs}], & \rho_a \le \rho \le \rho_{max} \end{cases}. \tag{7.4}$$

For all results reported here we set the absorber domain of redial width $\rho_{abs} = 24$ a.u., exceeding well the field induced electron oscillation $\alpha_d = E_0/\omega^2$ of the electron. The physical domain has the redial width of $\rho_a = \rho_{max} - \rho_{abs} = 104$ a.u..

An energy analysis of the ionization probabilities is employed based on a Fourier analysis of the associated flux (electronic current density) to describe MPADs [22, 23] after integrating out z perpendicular to the molecular (x, y) and laser polarization plane, i.e., averaged over z. For attosecond photoionization processes, the laser duration T is very short. At a large asymptotic point ρ_0, the angular flux distributions can be ignore, i.e., $1/\rho_0 \partial/\partial\theta|_{\rho_0}\hat{e}_\theta \ll \partial/\partial\rho|_{\rho_0}\hat{e}_\rho$. The time-independent energy-resolved angular differential yield (photoelectron spectra) is obtained by a Fourier transform:

$$\Psi(\theta, E_e)|_{\rho_0} = \int_{t_p}^{\infty} \psi(\theta, t)|_{\rho_0} e^{iE_e t} dt,$$

$$\Psi'(\theta, E_e)|_{\rho_0} = \int_{t_p}^{\infty} \frac{\partial\psi(\theta, t)}{\partial\rho}|_{\rho_0} e^{iE_e t} dt, \tag{7.5}$$

$$\mathcal{J}(\theta, E_e) \sim \Re\left[\frac{1}{2i}\Psi'^*(\theta, E_e)|_{\rho_0}\Psi(\theta, E_e)|_{\rho_0}\right],$$

where t_p is time after pulses switch off. The radial flux is calculated at $\rho_0 = 100$ a.u.. $E_e = p_e^2/2$ is the kinetic energy of an ionized electron with wave vector $k = p_e = 2\pi/\lambda_e$ (in a.u.), $p_e = (p_x^2 + p_y^2)^{1/2}$ is the momentum of a photoelectron of wavelength λ_e.

Since only the radial electronic flux is taken into account in simulations, we may define $\theta_{pR} = \theta$ where θ_{pR} is the angle between the electron momentum $\mathbf{p}$ and the molecular axis $\mathbf{R}$. With the transformation $p_x = p_e \cos\theta_{pR}$ and $p_y = p_e \sin\theta_{pR}$, we then obtain the 2D momentum distributions of a photoelectron from (7.5). The MPAD at a specific kinetic energy E_{en} is obtained by integrating over energy,

$$\mathcal{J}^{E_{en}}(\theta_{pR}) = \int_{E_{en}-\omega/2}^{E_{en}+\omega/2} dE_e \mathcal{J}(\theta_{pR}, E_e). \tag{7.6}$$

7.3 Electron Interference in Attosecond Photoionization

It has been shown that the directly ejected photoelectron with energies $\omega - I_p$ (momentum $p^r = \sqrt{2(\omega - I_p)}$) released from the two centers induces laser induced electron diffraction (LIED) by interference in the continuum [67, 68].

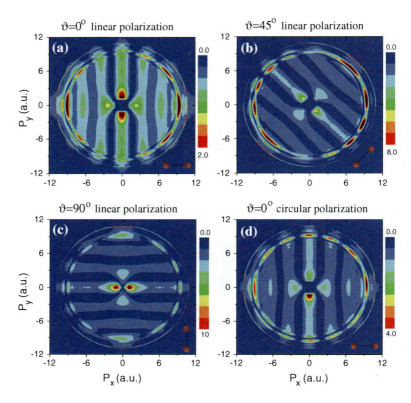

Fig. 7.1 Photoionization momentum distributions (p_x, p_y) in **a–c** *linear* and **d** *circular* polarization ionization for H_2^+ at $R_e = 2$ a.u. by $\lambda = 1$ nm laser pulses with $I_0 = 1.0 \times 10^{14}$ W/cm^2 and $T = 33$ as. The molecular orientation angles between the molecular axis and the x axis are **a, d** $\vartheta = 0°$, **b** $45°$, and **c** $90°$

(For distinguishment, we define the momenta of the directly ejected electron and the scattering electron as p^r and p^s, respectively). Such LIED fringes depend on the symmetry of the initial electronic state. It has been suggested that the ionized electron wave packets are elastically scattered by ions due to ultrashort wavelength of pulses [69]. The interference between the two scattering electron wave packets in the continuum lead to a Young-type interference [70].

Figure 7.1 shows photoelectron momentum (p_x, p_y) distributions of the H_2^+ ground $1s\sigma_g$ electronic state at equilibrium $R_e = 2$ a.u. with different angles $\vartheta = 0°$, $45°$, and $90°$, by intense ultrashort linearly and circularly polarized attosecond $\lambda = 1$ nm x-ray pulses, propagating along the z axis. Pulse intensities and durations are fixed at $I_0 = 1.0 \times 10^{14}$ W/cm^2 ($E_0 = 0.053$ a.u.) and $T = 10\tau = 33$ as. It is found that both ring structures with angle nodes and pronounced stripes are produced. Momentum rings are shown to be sensitive to the laser polarization and the molecular orientation angle ϑ. Of note is that the momentum stripes are always perpendicular to the molecular axis. In Fig. 7.1 radii of the momentum rings are $p^r = 9.43$ a.u.

7 Coherent Electron Wave Packet

corresponding to $p^r = \sqrt{2(\omega - I_p)}$, where $I_p = 1.1$ a.u. is the ionization potential and stripe momenta are $p^s_\perp = 2n\pi/R_e = n\pi$, $n = 0, \pm1, \pm2, \cdots$, with interval $\Delta p^s = 3.14 = \pi$ a.u.. The stripe momenta $p^s = \sqrt{p^{s2}_\parallel + p^{s2}_\perp} < p^r$, where $p^s_\parallel$ is the stripe momentum parallel to the molecular axis and p^r is the momentum of ejected electrons. Due to short duration of the pulses, momentum distributions exhibit slight asymmetry. Moreover multiple rings are also observed because of interference of ejected electron wave packets. The occurrence of the ring distributions with momentum $p^r = 9.43$ a.u. in Fig. 7.1 mainly results from LIED by the x-ray pulse, e.g. [22, 23] whereas the momentum stripes comes from interference of the coherent continuum scattering electron wave packets.

Assuming the equal two center scattering amplitude $f(\Omega)$, the wavefunction of the scattering electron from the two nuclear centers $\pm\mathbf{R}/2$ can be given

$$\psi_{sc}(\mathbf{r}) \sim f(\Omega) \left(\frac{e^{ip|\mathbf{r}+\mathbf{R}/2|}}{|\mathbf{r}+\mathbf{R}/2|} + \frac{e^{ip|\mathbf{r}-\mathbf{R}/2|}}{|\mathbf{r}-\mathbf{R}/2|} \right), \tag{7.7}$$

where $\mathbf{r}$ is the position of the scattering electron with respect to the molecular center. The asymptotic form of the wavefunction as $r \to \infty$ gives the amplitude of the scattering electron wave

$$\psi_{sc}(\mathbf{r}) \sim f(\Omega) \frac{e^{ipr}}{r} (e^{i\mathbf{p}^s \cdot \mathbf{R}/2} + e^{-i\mathbf{p}^s \cdot \mathbf{R}/2}), \tag{7.8}$$

where the vector $\mathbf{p}^s = p\mathbf{r}/r$ is the oriented along the direction of the scattering electron. We then can reduce the effective photoelectron scattering amplitude,

$$\mathscr{F}(\mathbf{p}^s) \sim |1 + \cos(\mathbf{p}^s \cdot \mathbf{R})|^2 \tag{7.9}$$

From (7.9) we see that the ionization amplitude of the scattering electron follows a cosine interference.

We display the H_2^+ MPADs of the scattering photoelectron with for example momenta $p^s = 6.28$ a.u. and wavelengths $\lambda^s_e = 1$ a.u. at $R_e = 2$ a.u. in Fig. 7.2a and $p^s = 3.59$ a.u. and $\lambda^s_e = 1.75$ a.u. at $R_c = 7$ a.u. in Fig. 7.2b in the $\lambda = 1$ nm circularly polarized attosecond x-ray laser pulse. From (7.9), we get that for H_2^+ at equilibrium R_e and the critical distance R_c with momenta $p^s = 6.28$ a.u. and 3.59 a.u., the corresponding angles between the photoelectron momentum and the molecular axis are $\theta_{p^s R_e} = \cos^{-1}(2n\pi/p^s R_e) = \cos^{-1}(n/2) = 0°, \pm60°, \pm90°, \pm120°$, and $180°$ and $\theta_{p^s R_c} = \cos^{-1}(2n\pi/p^s R_c) = \cos^{-1}(n/4) = 0°, \pm41°, \pm60°, \pm75° \pm90°, \pm105°, \pm120°, \pm139°$ and $180°$. These values also agree well with the numerical results in Fig. 7.2.

In Fig. 7.1 there are four humps around the molecular center in the low momentum region, which come from the scattering electrons of the molecule H_2^+ with the 1 nm x-ray laser pulse. Due to effects of the nonspherical asymmetric molecular Coulomb potential, the ionization rate is a function of the molecular orientation [71]. As a result

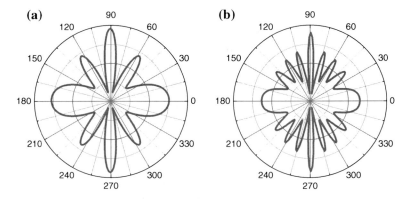

Fig. 7.2 MPADs in *circular* polarization ionization of $x-$ oriented H_2^+ at **a** equilibrium $R_e = 2$ a.u. and **b** critical internuclear distance $R_c = 7$ a.u. by the $\lambda = 1$ nm laser pulses with $I_0 = 1.0 \times 10^{14}$ W/cm^2 and $T = 33$ as for scattering electrons. The corresponding electron momenta and wavelengths are **a** $p^s = 2\pi = 6.28$ a.u. and $\lambda_e = 1.0$ a.u., and **b** $p^s = 8\pi/7 = 3.59$ a.u. and $\lambda_e = 1.75$ a.u.

the perpendicular (to the molecular axis) ionization distributions dominate. Of note is that for such a few cycle 1 nm x-ray laser pulse, a red shift of the photoelectronic spectra is also induced. Since the ionization distribution is functions of the pulse spectral width and diminishing electronic Franck-Condon factors, the momenta of these humps are slightly less than the the theoretical predictions. This energy shift has also been observed in attosecond photoionization processes of H and He atoms with linearly polarized ultrashort few cycle XUV laser pulses [72, 73]. Due to Coulomb attraction, the red shift is also sensitive to the photoelectron kinetic energy, leading to a pronounced "breaking" of the energy conservation in the low momentum region.

We next present results for a simple equilateral multi-center molecular ion H_3^{2+} in D_{3h} symmetry. A previous 2D model was used to study EI for this nonlinear system [74]. Figure 7.3 shows the photoelectron momentum distributions of the symmetric stretched H_3^{2+} ground A_1' electronic state at $R = 4$ a.u. in both linearly and circularly polarized $\lambda = 1$ nm attosecond x-ray laser pulses. The laser parameters are the same as those used in Fig. 7.1 for H_2^+. We note that in the H_3^{2+} ionization processes both the ring and stripe structures are also exhibited. For this nonliear multi-center photoionization, the interference is simply the sum of the contributions from all centers. The ionized electron wave packets from different centers will interfere with each other. Therefore, a set of alternating bright and dark interference fringes with spots instead of stripes are obtained. Since the angle between the molecular axes as well as the corresponding interference stripes of the scattering electron is $\pi/3$, the MPADs can be written as [63]:

$$\mathscr{F}(\mathbf{p}) \sim |\cos[pR\cos(\theta_{pR})/2]|^2 + |\cos[pR\cos(\theta_{pR} + \pi/3)/2]|^2$$
$$+ |\cos[pR\cos(\theta_{pR} + 2\pi/3)/2]|^2 \qquad (7.10)$$

7 Coherent Electron Wave Packet

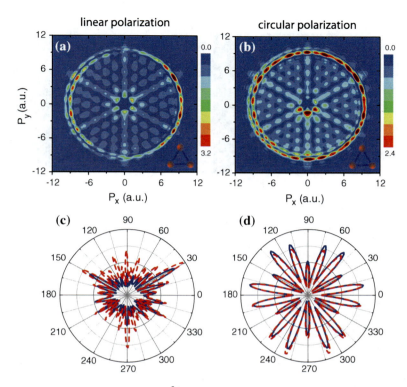

Fig. 7.3 **a** Triangular, D_{3h} symmetry H_3^{2+} photoionization momentum distributions (p_x, p_y) at $R = 4$ a.u. in **a** *linear* and **b** *circular* polarization ionization by $\lambda = 1$ nm attosecond x-ray laser pulses at $I_0 = 1.0 \times 10^{14}$ W/cm^2 and $T = 33$ as. **c** *Ring* and **d** *stripe* electron angular distributions for linear (*solid lines*) and circular (*dashed lines*) polarizations. The corresponding momenta and wavelengths are $p^r = 9.43$ a.u., $\lambda_e^r = 0.666$ a.u. and $p^s = 3\pi/2 = 4.71$ a.u., $\lambda_e^s = 1.33$ a.u.

where $p = p^{r/s}$ for scattered electrons and θ_{pR} is the angle between the electron momentum **p** and the molecular internuclear axis **R** along the x axis. We then get the momentum interval between the neighboring interference spots $\Delta p^s = \Delta p / \sin \pi/3 = \pi/\sqrt{3}$ a.u..

The angular distributions of the scattering electrons in H_3^{2+} are also presented in Fig. 7.3c and d. The ring electron momenta and wavelengths are chosen as $p^s = 3\pi/2 = 4.71$ a.u. and $\lambda_e^s = 1.33$ a.u.. Figure 7.3c show that MPADs with momentum $p^r = 9.43$ a.u. and wavelength $\lambda_e^r = 0.666$ a.u. are critically sensitive to the laser polarization. For the multi-center LIED patterns with linear and circular polarizations, the emission angles of the ionized electron are the same, but with different amplitudes. This difference is mainly due to the effects of the orientation and polarization dependent ionization rate. However in Fig. 7.3d for the sctripe electron, near identical MPADs are obtained in both linearly and circularly polarized laser pulses. In the elastic scattering processes of an electron, each of the multiple centers is an equivalent source of the waves and thus the interference between these

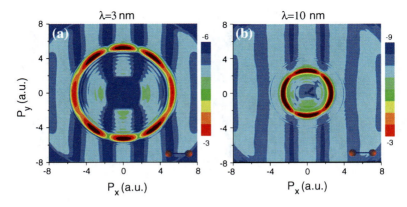

Fig. 7.4 Photoionization momentum distributions (p_x, p_y) in *circular* polarization ionization of $x-$ oriented H_2^+ at $R_e = 2$ a.u. by $I_0 = 1.0 \times 10^{14}$ W/cm^2 attosecond laser pulses at **a** $\lambda = 3$ nm and $T = 99$ as and **b** $\lambda = 10$ nm and $T = 330$ as. (Contours are equidistant on a logarithmic scale)

scattering electron waves produces polarization independent MPADs. It should be noted that in the low energy region, MPADs also depend on laser polarization due to nonspherical Coulomb potential effects. Comparison of Fig. 7.1a and d also shows laser polarization independence of interference stripes in the H_2^+ ionization processes.

We finally show interference and diffraction patterns in longer wavelength XUV laser fields. Figure 7.4 illustrates photoelectron momentum (p_x, p_y) distributions for x oriented H_2^+ at equilibrium with circularly polarized laser pulse at $\lambda = 3$ nm and $T = 99$ as in panel (a) and $\lambda = 10$ nm and $T = 330$ as in panel (b). The other parameters are the same as those used in Fig. 7.1d. It is found that both momentum rings and stripes are produced. Of note is that even in the region of momentum bigger than $\sqrt{2(\omega - I_p)}$, stripes are also observed. At $\lambda = 3$ nm diffraction patterns with eight angle nodes in momentum ring with radii $p^r = 5.48$ a.u. are induced whereas at $\lambda = 10$ nm no diffraction is possible since $\lambda_e^r = 2.63$ a.u. $> R_e$. The rings are coherent superposition of the ejected electron wave packets from the two centers. For both cases, the stripe momentum interval is always fixed at π a.u., which is consistent with results for $\lambda = 1$ nm in Fig. 7.1 and predictions in (7.9). Such stripe fringe in fact results from interference between the coherent scattering electron wave packets in higher order above threshold ionization after absorption of multiple photons, i.e., $n\omega - I_p$ but with weak signal intensity.

7.4 Electron Interference in Multiple Pathway Ionization

In the previous sections we investigated attosecond photoionization with single color circularly polarized XUV laser pulses. Two color XUV laser pulses have been used to control photoionization processes in atomic and molecular systems by interference

7 Coherent Electron Wave Packet

effects of CEWPs [75–79]. The breaking of spatial symmetry in MPADs, arising from simultaneous interference photoionization processes was reported in both atomic Rb and molecular NO [75]. Based on a quantum interference between two photoionization processes, a method for measurement of phase differences of continuum electron wave functions was provided in both linearly and elliptically polarized laser fields [78]. We show that the relative carrier envelop phase (CEP) ϕ dependent asymmetry in MPADs in MATI is due to the interference effect of CEWPs from different multi-photon ionization pathways.

7.4.1 Asymmetry of MPADs

We adopt the two color circularly polarized laser pulses, $\mathbf{E}_1(t)$ and $\mathbf{E}_2(t)$ with the form

$$\mathbf{E}_1(t) + \mathbf{E}_2(t) = \mathbf{e}_x f(t)[E_1 \cos(\omega_1 t + \phi_1) + E_2 \cos(\omega_2 t + \phi_2)] \\ + \mathbf{e}_y f(t)[E_1 \sin(\omega_1 t + \phi_1) + E_2 \sin(\omega_2 t + \phi_2)], \qquad (7.11)$$

at wavelengths $\lambda_1 = 60$ nm and $\lambda_2 = 30$ nm, corresponding to angular frequencies $\omega_1 = 0.76$ a.u. $= 20.7$ eV and $\omega_2 = 2\omega_1 = 1.52$ a.u. $= 41.6$ eV, which are respectively below and above the equilibrium ($R_e = 2$ a.u.) ionization potential to ionize H_2^+, and phases ϕ_1 and ϕ_2. It should be noted that with the $\lambda_1 = 60$ nm laser pulse, no resonance excitation occurs, thus avoiding additional unexpected strong field phenomena due to effects of electronically excited intermediate states. Such a laser pulse can produce photoelectron wave packets with the same kinetic energies, and steers them through different pathways in the continuum, thus creating the interference effects of CEWPs in the MPAD spectra. The molecular ion H_2^+ is pre-oriented before ionization. The intensities of the driving laser pulses are also fixed at $I_1 = 2I_2 = 1.0 \times 10^{13}$ W/cm^2 ($E_1 = \sqrt{2}E_2 = 1.69 \times 10^{-2}$ a.u. $= 8.68 \times 10^7$ V/cm). At the intensity $I_1 = 1.0 \times 10^{13}$ W/cm^2 and frequency $\omega_1 = 0.76$ a.u., the corresponding ponderomotive energy $U_p = I_1/4\omega_1^2 = 1.2 \times 10^{-4}$ a.u., corresponding to a Keldysh parameter $\gamma = \sqrt{I_p/2U_p} = 67.7 \gg 1$ and thus implying a multi-photon regime [64] as well.

We consider the photoionization by two color circularly polarized laser pulses with duration $T = 12\tau = 50$ a.u. $= 1.2$ fs (1 optical cycle, o.c., $\tau = 2\pi/\omega_2 = 100$ as) for each XUV laser pulse, i.e., $T = 6$ o.c. for $\lambda_1 = 60$ nm and $T = 12$ o.c. for $\lambda_2 = 30$ nm laser pulses. Figure 7.5 shows the 2D momentum (p_x, p_y) distributions of x oriented H_2^+ ($\vartheta = 0°$) for different relative CEPs ϕ of the circularly polarized two color XUV laser pulses. The fundamental XUV laser pulse wavelength and CEP are fixed at $\lambda_1 = 60$ nm and $\phi_1 = 0$, which serves as a reference, while the second $\lambda_2 = 30$ nm XUV laser pulse has variable peak time and phase ϕ_2 i.e., the relative CEP $\phi = \phi_2$. In the momentum (p_x, p_y) distributions, the signature of scattering effects appears as nodes or "wings", which are normally independent of CEP ϕ for single color

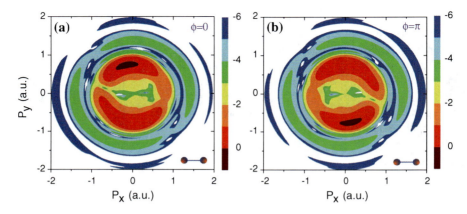

Fig. 7.5 2D momentum distributions (p_x, p_y) of photoelectron spectra in *circular* polarization ionization of x-oriented $\vartheta = 0°$ H_2^+ at equilibrium $R_e = 2$ a.u. in the two color $\lambda_1 = 60$ nm $E_1(t)$ and $\lambda_2 = 30$ nm $E_2(t)$ XUV laser pulses with duration $T = 12\tau = 1.2$ fs, intensities $I_1 = 2I_2 = 1.0 \times 10^{13}$ W/cm^2, and different CEPs **a** $\phi_1 = \phi_2 = 0$, i.e., relative pahse $\phi = 0$ and **b** $\phi_1 = 0$ and $\phi_2 = \pi$, i.e., $\phi = \pi$. (• − • corresponds to the molecular axis)

multiple cycle laser pulses [12]. The asymmetry in positive-negative momentum distributions strongly depends on the relative CEP ϕ, reflecting interference effects of coherent electron wave packets created respectively by the two laser pulses. The photoelectron momentum distributions display three MATI rings at $p_{e1} = 0.775$ a.u. ($E_{e1} = 0.3$ a.u.), $p_{e2} = 1.456$ a.u. ($E_{e2} = 1.06$ a.u.), and $p_{e3} = 1.908$ a.u. ($E_{e3} = 1.82$ a.u.), with the corresponding photoelectron wavelengths $\lambda_{e1} = 8.1$ a.u., $\lambda_{e2} = 4.3$ a.u., and $\lambda_{e3} = 3.8$ a.u., i.e., the first three MATI peaks separated by one photon energies $\Delta E = 0.76$ a.u. $= \omega_1$, which are asymmetric and mirror image of each other. Moreover since $\lambda_e > R$, no LIED is possible and in the momentum distributions angle nodes mainly localize along the x and y polarization directions.

The asymmetry introduced by the relative CEP of the two color circularly polarized XUV laser pulse appears much clearer in the MATI angular distribution. Figure 7.6 shows the corresponding MATI angular distributions $\mathcal{J}^{E_e}(\theta)$ at the three special photoelectron kinetic energies $E_{e1} = 0.3$ a.u., $E_{e2} = 1.06$ a.u., and $E_{e3} = 1.82$ a.u.. The results show that the asymmetry is also critically sensitive to the photoelectron kinetic energy E_e (momentum p_e). As illustrated in Fig. 7.6a at $E_{e1} = 0.3$ a.u., a positive-negative asymmetry appears in the MPADs where the perpendicular y distributions dominates. At $E_{e2} = 1.06$ a.u., the amplitude of parallel distributions increases, and at $E_{e3} = 1.82$ a.u., the parallel x distributions dominates. Of note is that as the photoelectron kinetic energies E_e increase, the forward-backward asymmetry in the parallel x distributions increases whereas the up-down asymmetry in the MPADs are nearly constant. The dependence of photoelectron kinetic energy E_e on the asymmetry in MATI angular distributions mainly indicates that the interference is also a function of E_e.

7 Coherent Electron Wave Packet 161

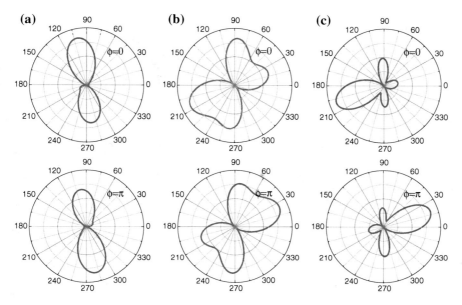

Fig. 7.6 MPADs $\mathscr{J}^{E_e}(\theta)$ at different MATI photoelectron kinetic energies **a** $E_{e1} = 2\omega_1 - I_p = 0.3$ a.u., **b** $E_{e2} = 3\omega_1 - I_p = 1.06$ a.u., and **c** $E_{e3} = 4\omega_1 - I_p = 1.82$ a.u. in *circular* polarization ionization of x-oriented $\vartheta = 0°$ H_2^+ at equilibrium $R_e = 2$ a.u. in the two color $\lambda_1 = 60$ nm $\mathbf{E}_1(t)$ and $\lambda_2 = 30$ nm $\mathbf{E}_2(t)$ XUV laser pulses with duration $T = 12\tau = 1.2$ fs, intensities $I_1 = 2I_2 = 1.0 \times 10^{13}$ W/cm² and different CEPs $\phi_1 = \phi_2 = 0$, i.e., $\phi = 0$ (*upper row*) and $\phi_1 = 0$ and $\phi_2 = \pi$, i.e., $\phi = \pi$ (*bottom row*), corresponding to Fig. 7.5

We next show effects of the molecular orientation on the asymmetry of MPADs. Figure 7.7 displays results of the molecular photoelectron momentum (p_x, p_y) distributions for an oriented $\vartheta = 45°$ molecular ion H_2^+. The same laser pulses are used as in Fig. 7.5. Again a clear asymmetry is observed in the momentum distributions in MATI. Comparison of Figs. 7.5 and 7.7 shows that the asymmetry is essentially sensitive to the molecular orientation angle ϑ. It is mainly perpendicular to the molecular axis (• − •) in both cases. As we discuss bellow in Sect. 7.4.2, in the molecular photoionization processes, the asymmetry of the MPADs in MATI comes from the resulting nonsymmetric electric fields of the two superposed laser pulses with different CEPs. Moreover due to the molecular nonspherical asymmetric Coulomb potential the distributions of the photoelectron depend on the laser polarization directions. Consequently, the positive-negative asymmetry see in Figs. 7.5 and 7.7 can always be produced and steered by the relative CEPs of the two color XUV laser pulses. This is an important aspect of two color circularly polarized attosecond pulses. The effects of electron interference are dependent of molecular orientation in circular polarization ionization as well. The requirement of molecular orientation can be readily achieved with recent laser orientation technology [80].

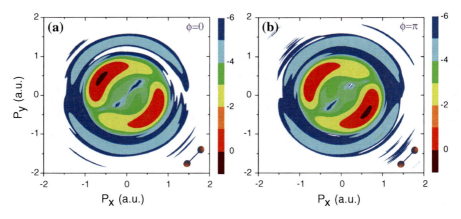

Fig. 7.7 2D momentum distributions (p_x, p_y) of photoelectron spectra in *circular* polarization ionization of the $\vartheta = 45°$ oriented molecular ion H_2^+ at equilibrium $R_e = 2$ a.u. in the two color $\lambda_1 = 60$ nm $\mathbf{E}_1(t)$ and $\lambda_2 = 30$ nm $\mathbf{E}_2(t)$ XUV laser pulses with duration $T = 12\tau = 1.2$ fs, intensities $I_1 = 2I_2 = 1.0 \times 10^{13}$ W/cm², and different CEPs **a** $\phi_1 = \phi_2 = 0$, i.e., relative phase $\phi = 0$ and **b** $\phi_1 = 0$ and $\phi_2 = \pi$, i.e., $\phi = \pi$. (•−−• corresponds to the molecular axis)

7.4.2 Description of Multiple Pathway CEWP Interference

We adopt a perturbative theoretical model of multi-photon ionization to understand the sensitivity of the CEP ϕ to the asymmetry in the photoionization momentum and angular distributions in Figs. 7.5 and 7.6, as shown in Appendix 1. For the numerical results illustrated in Figs. 7.5 and 7.6, the total phase difference of the driving laser pulses is defined as $\phi = \phi_2 - \phi_1 = \phi_2$ with $\phi_1 = 0$, simplifying the phase difference $\Delta\phi = \phi$ in the perturbation model in Appendix 1. As a result, in (7.23)/(7.24), (7.26)/(7.27) and (7.33)/(7.34) for the total MPADs, the CEP dependent asymmetry terms can be simply expressed as

$$\sigma^s \propto \cos(\Delta\phi + \Delta\xi) = \cos(\phi + \Delta\xi). \tag{7.12}$$

The continuum electron wave function phase difference $\Delta\xi$ is insensitive to the laser pulse CEP ϕ in the perturbative limit. Then the CEP ϕ dependence of angular distributions becomes a function of $\cos\phi$. The relation between asymmetric MPADs at $\phi = 0$ and π is simply,

$$\sigma^s(\phi = 0) = -\sigma^s(\phi = \pi) \propto \cos(\Delta\xi). \tag{7.13}$$

From (7.13) we see that the MPADs for CEPs $\phi = 0$ and π do exhibit an exact opposite asymmetry. Therefore as shown in Figs. 7.5 and 7.6 the momentum and angular distributions of MATI spectra are positive-negative asymmetric in the momentum plane and mirror images of each other.

7 Coherent Electron Wave Packet 163

Furthermore we see, in (7.23)/(7.24), (7.26)/(7.27) and (7.33)/ (7.34), that the asymmetry terms are also sensitive to photoelectron MATI kinetic energies E_{en}, i.e., for $E_{e1} = 2\omega_1 - I_p$, $d\sigma(E_{e1})/d\Omega \propto \cos^3\theta$ and $\sin^3\theta$, for $E_{e2} = 3\omega_1 - I_p$, $d\sigma(E_{e2})/d\Omega \propto \cos^5\theta$ and $\sin^5\theta$, and for $E_{e3} = 4\omega_1 - I_p$, $d\sigma(E_{e3})/d\Omega \propto \cos^5\theta + \cos^7\theta$ and $\sin^5\theta + \sin^7\theta$. As shown in MATI spectra in Fig. 7.6a at E_{e1} the electron wave packet produced though the two-photon ($\omega_1 + \omega_1$) absorption interferes with that generated by the one-photon (ω_2) ionization, therefore leading to the simultaneous interference in the photoelectron spectra and opposite asymmetries for the phase difference $\phi = \pi$ in accordance with (7.23) and (7.24). In Fig. 7.6b at E_{e2} the interference effect again produces opposite asymmetries as a result of the two transition pathways, three-photon ($\omega_1 + \omega_1 + \omega_1$) and two-photon ($\omega_1 + \omega_2$) ionizations, and in Fig. 7.6c at E_{e3} of three channel interference, four-photon ($\omega_1 + \omega_1 + \omega_1 + \omega_1$), three-photon ($\omega_1 + \omega_1 + \omega_2$) and two-photon ($\omega_2 + \omega_2$) transitions. Of note is that the asymmetry in Fig. 7.6c for E_{e3} is stronger than those in Fig. 7.6a for E_{e1} and in Fig. 7.6b for E_{e2}. As shown in (7.33) and (7.34) three angle interference terms result in the asymmetry in this MPAD in Fig. 7.6c. We may thus attribute the enhancement of the MPAD asymmetry at the high energy E_{e3} to increase interference effects of multiple multi-photon ionization pathways.

Moreover, the time delay between the two color circularly polarized XUV laser pulses can influence the asymmetry of MPADs. As we discussed previously [28] varying the time delay corresponds to altering the CEP of the total laser fields. As a result the phase difference $\Delta\xi$ changes, leading to a periodical oscillation of the asymmetry. It should be noted that the longer the time delay, the decreasing of the overlap of the two color laser pulses produces less simultaneous interference amplitudes β in Appendix 1, that is the asymmetry decrease.

7.4.3 Influence of Pulse Intensity on the Asymmetry of MPADs

We study next the effects of the laser intensity on the molecular photoionization. Figures 7.8 and 7.9 display MATI angular distributions at $E_{e1} = 2\omega_1 - I_p = 0.3$ a.u., $E_{e2} = 3\omega_1 - I_p = 1.06$ a.u., and $E_{e3} = 4\omega_1 - I_p = 1.82$ a.u. for x-oriented $\vartheta = 0°$ H_2^+ at equilibrium $R_e = 2$ a.u. and photoelectron energies at two higher intensities $I_1 = 2I_2 = 7.5 \times 10^{13}$ W/cm^2 and 5.0×10^{14} W/cm^2 of the two color circularly polarized laser pulses. The other laser parameters are the same as those used in Figs. 7.5 and 7.6. The CEP dependent asymmetry is obtained again. It is also found that the asymmetry is critically sensitive to pulse intensities.

We turn to the perturbative multi-photon ionization model for an n-photon ionization process. The transition matrix element $\mathscr{A}_{n\omega}$ can also be simply written as

$$\mathscr{A}_{n\omega} \propto \mathscr{D}^{(n)}(E)^n \tag{7.14}$$

where $\mathscr{D}^{(n)}$ is the n-photon ionization amplitude,

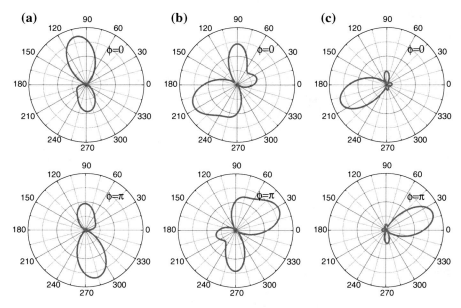

Fig. 7.8 MPADs $\mathscr{J}^{E_e}(\theta)$ at specific MATI photoelectron kinetic energies **a** $E_{e1} = 2\omega_1 - I_p = 0.3$ a.u., **b** $E_{e2} = 3\omega_1 - I_p = 1.06$ a.u., and **c** $E_{e3} = 4\omega_1 - I_p = 1.82$ a.u. in *circular* polarization ionization of x-oriented $\vartheta = 0°$ H_2^+ at equilibrium $R_e = 2$ a.u. in the two color $\lambda_1 = 60$ nm $\mathbf{E}_1(t)$ and $\lambda_2 = 30$ nm $\mathbf{E}_2(t)$ XUV laser pulses with duration $T = 12\tau = 1.2$ fs, intensities $I_1 = 2I_2 = 7.5 \times 10^{13}$ W/cm^2 and different CEPs $\phi_1 = \phi_2 = 0$, i.e., relative phase $\phi = 0$ (*upper row*) and $\phi_1 = 0$ and $\phi_2 = \pi$, i.e., $\phi = \pi$ (*bottom row*)

$$\mathscr{D}^{(n)} \sim \langle \psi_c^{E_{en}}(\mathbf{r}) | \tilde{D}(\mathbf{r}) | \psi_0(\mathbf{r}) \rangle. \quad (7.15)$$

$\tilde{D}(\mathbf{r})$ is an nth order electric dipole operator corresponding to the transition from the initial state $\psi_0(\mathbf{r})$ to the continuum state $\psi_c^{E_{en}}(\mathbf{r})$. From (7.14) and (7.15) we see that for the two ω_1 photon process the pulse intensity dependent ionization amplitude is $\mathscr{A}_{2\omega_1} \sim \mathscr{D}^{(2)}(E_1)^2$ whereas for the single ω_2 photon process $\mathscr{A}_{\omega_2} \sim \mathscr{D}^{(1)}(E_2)$. Increase of pulse intensities E_1 and E_2 results in much more electrons after absorption of two ω_1 photon. Consequently, the CEWP interference between the two ionization pathways is enhanced, leading to increase of the asymmetry. The similar process occurs for the higher order multi-photon ionizations with energies E_{e2} and E_{e3}. Since the ionization amplitude $\mathscr{A}_{n\omega}$ is a function of photon numbers, $(E)^n$, at higher order photoelectron kinetic energies E_{en}, the multi-photon ionization process is enhanced as well with increase of the pulse intensity. Therefore stronger asymmetries are obtained.

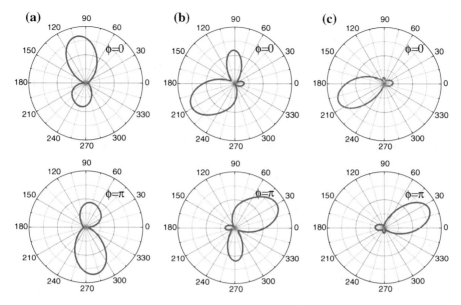

Fig. 7.9 MPADs $\mathscr{J}^{E_e}(\theta)$ at specific MATI photoelectron kinetic energies **a** $E_{e1} = 2\omega_1 - I_p = 0.3$ a.u., **b** $E_{e2} = 3\omega_1 - I_p = 1.06$ a.u., and **c** $E_{e3} = 4\omega_1 - I_p = 1.82$ a.u. in *circular* polarization ionization of x-orieted $\vartheta = 0°$ H_2^+ at equilibrium $R_e = 2$ a.u. in the two color $\lambda_1 = 60$ nm $\mathbf{E}_1(t)$ and $\lambda_2 = 30$ nm $\mathbf{E}_2(t)$ XUV laser pulses with duration $T = 12\tau = 1.2$ fs, intensities $I_1 = 2I_2 = 5.0 \times 10^{14}$ W/cm^2 and different CEPs $\phi_1 = \phi_2 = 0$, i.e., relative phase $\phi = 0$ (*upper row*) and $\phi_1 = 0$ and $\phi_2 = \pi$, i.e., $\phi = \pi$ (*bottom row*)

7.5 Conclusions

In summary we investigate electron interference in attosecond photoionization of molecules with circularly polarized XUV laser pulses from numerical solutions of the corresponding TDSEs for static nuclei. It is found that with intense attosecond XUV laser pulses, interference of scattering electron wave packets are produced. Striking interference patterns with stripes perpendicular to the molecular R axis with momentum interval $\Delta p^s = 2\pi/R$ are observed in molecular photoelectron momentum distributions for both H_2^+ and H_3^{2+}. This behavior is analogous to the Young's double slit interference experiment and has been interpreted by the interference model of scattering electron waves. Such interference stripes are shown to be insensitive to the laser polarization, confirming the prediction in (7.8) and (7.9). We also present interference effects of CEWPs from multiple pathway ionizations. We present MPADs in two color $\lambda_1 = 2\lambda_2$ circularly polarized XUV laser pulse. Photon energies ω_1 and ω_2 are respectively below and above the molecular ionization potential I_p, thus creating electron wave packets with same energies in the continuum by interfering ionization pathways. It is found that MATI angular distributions exhibit strong asymmetries, which are sensitive to the relative CEP of the two color driving

XUV circularly polarized laser pulses. This is attributed to the effects of interference between the coherent circular polarization electron wave packet created by multiple pathway ionizations. By adopting a perturbative multi-photon ionization model, we describe these features in the two color laser fields. Influence of the pulse intensity on the asymmetry in the MPADs are also presented. We show thus in general that attosecond pulses allow to probe symmetries of molecular states by instantaneous ionization at different molecular conformations.

Acknowledgments The authors thank RQCHP and Compute Canada for access to massively parallel computer clusters and CIPI (Canadian Institute for Photonic Innovations) for financial support of this research in its ultrafast science program. A. D. Bandrauk is also indebted to a Canada Research Chair for pursuing attosecond research.

Appendix 1

To understand the sensitivity of the asymmetry in the photoionization momentum and angular distributions to the relative phase ϕ, we adopt a perturbative theoretical model of a multi-photon ionization. Figure 7.10 illustrates the possible photoionization paths in the two color laser pulses. For the ionization at photoelectron kinetic energies $E_{e1} = 2\omega_1 - I_p$, the photoionization proceeds by two simultaneous processes, a direct one-photon transition at ω_2 and two one-photon transitions at $\omega_1 + \omega_1$. Usually for direct one-photon ionization processes by laser pulses, the ionization differential probability can be expressed simply in the dipole form [67]:

$$\frac{dP_{\text{ion}}}{d\Omega} \propto |\langle \psi_c^{E_e}(\mathbf{r})|T_H|\psi_0(\mathbf{r})\rangle|^2 = |\langle \psi_c^{E_e}(\mathbf{r})|\mathbf{r} \cdot \mathbf{E}|\psi_0(\mathbf{r})\rangle|^2, \qquad (7.16)$$

where $\psi_c^{E_e}(\mathbf{r})$ is the continuum electron wave function with energy E_e and $\psi_0(\mathbf{r})$ is the initial state. T_H is a transition operator corresponding to the transition from the initial state $|\psi_0\rangle$ to the continuum state $|\psi_c^{E_e}\rangle$. Under conditions $\lambda_e > R$ no diffraction of electron occurs. The circular polarization MPADs are simply superpositions of the linear parallel and perpendicular (to the molecular axis) polarization distributions [22, 23].

One-Color Pulse Ionization Processes

In a one-color laser pulse, the angular distributions of electrons emitted in one-photon ionization processes have the simple forms respectively $\mathscr{P}_\parallel(\omega_1) \propto \alpha_0 + \alpha_1 \cos^2\theta$ for the parallel x polarization, and $\mathscr{P}_\perp(\omega_1) \propto \alpha_0 + \alpha_1 \sin^2\theta$ for the perpendicular y polarization, where coefficients $\alpha_i, i = 0, 1, 2, \ldots$, depend on laser pulses and the initial state. Each transition matrix amplitude shows that the CEP ϕ has no influence on the multi-photon ionization angular distributions of the photoelectron spectra.

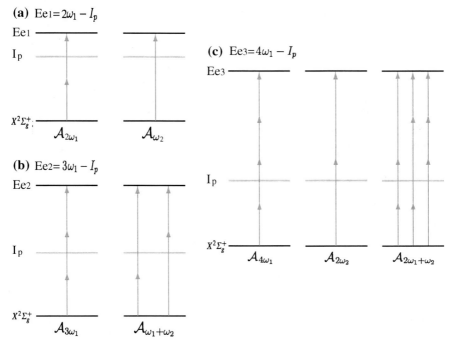

Fig. 7.10 Schematic illustrations of multi-photon ionization pathways from the initial H_2^+ $1s\sigma_g$ ground state at $R_e = 2$ a.u. in the ultrashort two color laser fields for the MATI peaks at energies **a** $E_{e1} = 2\omega_1 - I_p$, **b** $E_{e2} = 3\omega_1 - I_p$, and **c** $E_{e3} = 4\omega_1 - I_p$. $\mathscr{A}$ is the corresponding transition matrix element. *Short* and *long* arrows denote ω_1 and ω_2 photons ($\omega_2 = 2\omega_1$)

The two-photon ($\omega_1 + \omega_1$) transition matrix element $\mathscr{A}_{2\omega_1}$ reads in the perturbative dipole approximation as:

$$\begin{aligned}
\mathscr{A}_{2\omega_1} &= \langle \psi_c^{E_e} | T_H | \psi_0 \rangle \\
&= \int_{\lim \varepsilon \to 0} dE_n^e \frac{\langle \psi_c^{E_e} | \mathbf{r} \cdot \mathbf{E} | \psi_n \rangle \langle \psi_n | \mathbf{r} \cdot \mathbf{E} | \psi_0 \rangle}{E_0^e - E_n^e + \omega_1 + i\varepsilon} \\
&= \text{PP} \int dE_n^e \frac{\langle \psi_c^{E_e} | \mathbf{r} \cdot \mathbf{E} | \psi_n \rangle \langle \psi_n | \mathbf{r} \cdot \mathbf{E} | \psi_0 \rangle}{E_0^e - E_n^e + \omega_1} \\
&\quad - i\pi \delta(E_0^e - E_n^e + \omega_1) \langle \psi_c^{E_e} | \mathbf{r} \cdot \mathbf{E} | \psi_n \rangle \langle \psi_n | \mathbf{r} \cdot \mathbf{E} | \psi_0 \rangle \\
&= \text{Re}(\mathscr{A}_{2\omega_1}) + i\text{Im}(\mathscr{A}_{2\omega_1}),
\end{aligned} \quad (7.17)$$

where the integral sums over all intermediate (virtual) states $|\psi_n\rangle$. Equation (7.17) applies to continuum intermediate states $|\psi_n\rangle$ and/or high density Rydberg states [81]. The transition amplitude in (7.17) can thus be written as

$$\mathscr{A}_{2\omega_1} = \mathscr{R}_{2\omega_1} e^{i\eta_{2\omega_1}}, \tag{7.18}$$

$$\mathscr{R}_{2\omega_1} = \sqrt{[\text{Re}(\mathscr{A}_{2\omega_1})]^2 + [\text{Im}(\mathscr{A}_{2\omega_1})]^2},$$

$$\eta_{2\omega_1} = \tan^{-1}\left[\text{Im}(\mathscr{A}_{2\omega_1})/\text{Re}(\mathscr{A}_{2\omega_1})\right],$$

where $\eta_{2\omega_1}$ is the phase of the transition amplitude of the two ω_1 photon ionizations. The total transition matrix element is separated into a nonresonant (virtual) principle part PP integral and a resonant transition $\omega_1 = E_n^e - E_0^e$, where E_0^e and E_n^e is the energies of the initial and intermediate electronic states. For continuum energies or dense states such as Rydberg states, the PP integral becomes negligible due to cancellation from fluctuations of the denominators $E_0^e - E_n^e + \omega_1 > 0$ and $E_0^e - E_n^e + \omega_1 < 0$ in (7.17), i.e.

$$\text{PP} \int_{-\infty}^{\infty} dE_n^e \frac{\langle \psi_c^{E_e} | \mathbf{r} \cdot \mathbf{E} | \psi_n \rangle \langle \psi_n | \mathbf{r} \cdot \mathbf{E} | \psi_0 \rangle}{E_0^e - E_n^e + \omega_1} = 0. \tag{7.19}$$

The nonresonant (virtual) processes in (7.19) occur on a time scale $\tau_{nr} = 1/|E_0^e - E_n^e + \omega_1|$. In this limit can one assume that the total transition element depends on the resonant process, i.e. the imaginary fact of the transition amplitude in (7.17) [28, 81],

$$\mathscr{A}_{2\omega_1} \propto \langle \psi_c^{E_e} | \mathbf{r} \cdot \mathbf{E} | \psi_n \rangle \langle \psi_n | \mathbf{r} \cdot \mathbf{E} | \psi_0 \rangle \tag{7.20}$$

In polar coordinates the parallel/perpendicular dipole interaction simply takes the form $\mathbf{r} \cdot \mathbf{E} = r E (\mathbf{r}_0 \cdot \mathbf{e}_{x/y})$, where $\mathbf{r}_0$ is the unit position vector, and then the radial and angular variables in the transition matrix element are easily separated. The electric field is defined as $E(\omega_1) = E_0(\omega_1) \exp(i\phi_{\omega_1})$, where $E_0(\omega_1)$ is the electric strength and ϕ_{ω_1} is the pulse phase. The transition matrix element $\mathscr{A}_{2\omega_1}$ can then be written as [82]

$$\mathscr{A}_{2\omega_1} \propto T_{2\omega_1} f_{2\omega_1}(\theta) e^{i\xi_{2\omega_1}} E(\omega_1)^2$$
$$= T_{2\omega_1} f_{2\omega_1}(\theta) E_0(\omega_1)^2 e^{i\xi_{2\omega_1} + 2\phi_{\omega_1}}, \tag{7.21}$$

where $\xi_{2\omega_1} = \eta_{2\omega_1}$ is the continuum electron wave function phase, $\phi_{2\omega_1}$ and ϕ_{ω_1} are the pulse CEPs. $T_{2\omega_1}$ and $f_{2\omega_1}(\theta)$ are respectively the radial and angular parts of the reduced transition moments. In the case of bound state resonances only, $\text{Im}(\mathscr{A}_{2\omega_1}) = 0$ in (7.17), so the corresponding amplitude of $\text{Re}(\mathscr{A}_{2\omega_1})$ is the same as the continuum transition amplitude in (7.21) since the final state $|\psi_c\rangle$ is the same.

7 Coherent Electron Wave Packet

Two-Color Photoionization Processes

For a combination of bichromatic laser pulses of frequencies ω_1 and $\omega_2 = 2\omega_1$, the electron can be ionized via multiple pathways to reach the same final energies in the continuum simultaneously. We below consider respectively the photoionization at different kinetic energies by bichromatic laser pulses.

Interference of CEWPs at Energy $E_{e1} = 2\omega_1 - I_p$

For the MPADs with energy $E_{e1} = 2\omega_1 - I_p$, i.e., simultaneous two-photon ($\omega_1 + \omega_1$) in (7.20) and one-photon ($\omega_2 = 2\omega_1$) ionizations in (7.16), the total transition probability is the square of the two amplitudes with an interference term of the cross products of the two one- and two-photon ionization amplitudes, i.e., $d\sigma(E_{e1})/d\Omega = \mathscr{P}(2\omega_1) + \mathscr{P}(\omega_2) + \mathscr{P}(2\omega_1, \omega_2)$, where $\mathscr{P}(2\omega_1, \omega_2)$ is the interference term which can be simply written as [78, 79, 82]

$$\begin{aligned} \mathscr{P}(2\omega_1, \omega_2) &\propto \mathscr{A}^*_{2\omega_1} \cdot \mathscr{A}_{\omega_2} + \mathscr{A}_{2\omega_1} \cdot \mathscr{A}^*_{\omega_2} \\ &= T_{2\omega_1} f_{2\omega_1}(\theta) E_0(\omega_1)^2 E_0(\omega_2) \cos(\Delta\eta), \end{aligned} \tag{7.22}$$

where $\mathscr{A}_{2\omega_1}$ and $\mathscr{A}_{\omega_2}$ are respectively the matrix element of the two- and one-photon absorptions, and the angular factors are obtained in the perturbative limit $f_{2\omega_1}(\theta) = \langle(\mathbf{r}_0 \cdot \mathbf{e})^3\rangle$. The two field amplitudes are defined as $E(\omega_2) = E_0(\omega_2)\exp(i\phi_{\omega_2})$ and $E(\omega_1) = E_0(\omega_1)\exp(i\phi_{\omega_1})$, where $E_0(\omega_{1,2})$ is the electric strength and $\phi_{\omega_{1,2}}$ is the pulse phase. The total phase difference $\Delta\eta$ between the transition amplitudes for the two-pathway ionizations is the sum of difference phases of the laser pulses $\Delta\phi$ and of the continuum electron wave functions $\Delta\xi$, i.e., $\Delta\eta = \Delta\phi + \Delta\xi$ with $\Delta\phi = \phi_{\omega_2} - 2\phi_{\omega_1}$ and $\Delta\xi = \xi_{\omega_2} - \xi_{2\omega_1}$, where ξ_{ω_2} and $\xi_{2\omega_1}$ are respectively the phases of continuum electron wave functions for direct one ω_2 and two ω_1 photon ionizations. Then the total parallel MPADs can be finally written as sums of direct and interfering photoionization distributions,

$$\frac{d\sigma_\parallel(E_{e1})}{d\Omega} \propto \alpha_0 + \alpha_1 \cos^2\theta + \alpha_2 \cos^4\theta + \beta\cos^3\theta\cos(\Delta\eta). \tag{7.23}$$

The coefficient β is determined by the intensities of the two laser pulses. Similarly, the total perpendicular MPADs can be written as

$$\frac{d\sigma_\perp(E_{e1})}{d\Omega} \propto \alpha_0 + \alpha_1 \sin^2\theta + \alpha_2 \sin^4\theta + \beta\sin^3\theta\cos(\Delta\eta). \tag{7.24}$$

We note that (7.23) and (7.24) contain odd order terms in both $\cos\theta/\sin\theta$ and $\cos(\Delta\eta) = \cos(\Delta\phi + \Delta\xi)$. The simultaneous interference effect will therefore break the symmetry of the electron flux in forward/upward and backward/downward

170   K.-J. Yuan and A. D. Bandrauk

directions and the MPADs will vary periodically with the phase difference $\Delta\phi$ of the laser pulses. At $\Delta\eta = \pi/2$ there is no interference and at $\Delta\eta = 0$ and π one gets maximum asymmetry.

Interference of CEWPs at Energy $E_{e2} = 3\omega_1 - I_p$

Photoionization processes can also occur at higher kinetic energy $E_{e2} = 3\omega_1 - I_p$, as schematically illustrated in Fig. 7.10b, where electron ionizes through channels after absorption of three ω_1 photons ($\omega_1 + \omega_1 + \omega_1$), giving a direct transition $\mathscr{P}(3\omega_1)$ a one ω_1 and one ω_2 photon $\mathscr{P}(\omega_1 + \omega_2)$ transition. Therefore the simultaneous interference term between these two multi-photon ionization pathways can be written as

$$\mathscr{P}(3\omega_1, \omega_1 + \omega_2) \propto \mathscr{A}^*_{3\omega_1} \cdot \mathscr{A}_{\omega_1 + \omega_2} + \mathscr{A}_{3\omega_1} \cdot \mathscr{A}^*_{\omega_1 + \omega_2}$$
$$= T_{3\omega_1} f_{3\omega_1}(\theta) E_0(\omega_1)^4 E_0(\omega_2) \cos(\Delta\phi). \quad (7.25)$$

$\mathscr{A}_{3\omega_1}$ and $\mathscr{A}_{\omega_1 + \omega_2}$ are transition amplitudes from $\omega_1 + \omega_1 + \omega_1$ and $\omega_1 + \omega_2$ pathway ionizations. The corresponding angular factor is given by $f_{3\omega_1}(\theta) = \langle(\mathbf{r}_0 \cdot \mathbf{e})^5\rangle$. Then one can express the overall parallel polarization angular distributions

$$\frac{d\sigma_\parallel(E_{e2})}{d\Omega} \propto \alpha_0 + \alpha_1 \cos^2\theta + \alpha_2 \cos^4\theta + \alpha_3 \cos^6\theta$$
$$+ \beta \cos^5\theta \cos(\Delta\eta), \quad (7.26)$$

For the perpendicular y polarization angular distributions, the corresponding expression is

$$\frac{d\sigma_\perp(E_{e2})}{d\Omega} \propto \alpha_0 + \alpha_1 \sin^2\theta + \alpha_2 \sin^4\theta + \alpha_3 \sin^6\theta$$
$$+ \beta \sin^5\theta \cos(\Delta\eta), \quad (7.27)$$

where α_i and β are pulse dependent coefficients with transition amplitude phase differences $\Delta\eta = \Delta\phi + \Delta\xi$, $\Delta\phi = \phi_{\omega_2} - 2\phi_{\omega_1}$ and $\Delta\xi = \xi_{\omega_1 + \omega_2} - \xi_{3\omega_1}$. $\xi_{3\omega_1}$ and $\xi_{\omega_1 + \omega_2}$ are the phases of the continuum electron wave functions for the $3\omega_1$ and $\omega_1 + \omega_2$ pathways ionizations. As illustrated in Fig. 7.10a, the ionization is in fact a three-pathway ionization process. The $\omega_1 + \omega_1 + \omega_1$ transition corresponds to three successive one photon transitions giving rise to a single transition amplitude $\mathscr{A}_{3\omega_1} = \mathscr{R}_{3\omega_1} e^{i\eta_{3\omega_1}}$ whereas the $\omega_1 + \omega_2$ transition corresponds to two amplitudes, $\mathscr{A}'_{\omega_1 + \omega_2}$ and $\mathscr{A}'_{\omega_2 + \omega_1}$ whose sum is defined as $\mathscr{A}_{\omega_1 + \omega_2} = \mathscr{R}_{\omega_1 + \omega_2} e^{i\eta_{\omega_1 + \omega_2}}$. Again, a phase difference dependent asymmetry in the MPADs is obtained due to the odd order terms in both $\cos\theta/\sin\theta$ and $\cos(\Delta\eta)$ in (7.26) and (7.27).

7 Coherent Electron Wave Packet

Interference of CEWPs at Energy $E_{e3} = 4\omega_1 - I_p$

Similar results for total MPADs can be obtained for the ionization processes with next higher kinetic energy $E_{e3} = 4\omega_1 - I_p$ via five-pathway transitions after direct absorptions of four ω_1 photons ($\omega_1 + \omega_1 + \omega_1 + \omega_1$), and two 2ω photons ($\omega_2 + \omega_2$), and one 2ω and two ω photons ($\omega_1 + \omega_1 + \omega_2$), as shown in Fig. 7.10c. For this multi-pathway ionization, the interference is simply the sum of the contributions from all pathways. From Fig. 7.10c we note that in the ionization processes three-pathway interference occurs, i.e., the four ω_1 photon ionization with transition $\mathscr{A}_{4\omega_1} = \mathscr{R}_{4\omega_1} e^{i\eta_{4\omega_1}}$, two ω_2 photon ionization with transition $\mathscr{A}_{2\omega_2} = \mathscr{R}_{2\omega_2} e^{i\eta_{2\omega_2}}$, and the three photon ionization processes $\mathscr{A}_{2\omega_1+\omega_2} = \mathscr{R}_{2\omega_1+\omega_2} e^{i\eta_{2\omega_1+\omega_2}}$ corresponding to three amplitudes $\mathscr{A}'_{2\omega_1+\omega_2}$, $\mathscr{A}'_{\omega_1+\omega_2+\omega_1}$, and $\mathscr{A}'_{\omega_2+2\omega_1}$ interfere with each other. Then we get respectively $\mathscr{A}_{4\omega_1}$ and $\mathscr{A}_{2\omega_2}$ interference,

$$\mathscr{P}(4\omega_1, 2\omega_2) \propto \mathscr{A}^*_{4\omega_1} \cdot \mathscr{A}_{2\omega_2} + \mathscr{A}_{4\omega_1} \cdot \mathscr{A}^*_{2\omega_2}$$
$$= T_{4\omega_1} f_{4\omega_1}(\theta) E(\omega_1)^4 E(\omega_2)^2 \cos(2\Delta\eta_0), \qquad (7.28)$$

$\mathscr{A}_{4\omega_1}$ and $\mathscr{A}_{2\omega_1+\omega_2}$ interference,

$$\mathscr{P}(4\omega_1, 2\omega_1 + \omega_2) \propto \mathscr{A}^*_{4\omega_1} \cdot \mathscr{A}_{2\omega_1+\omega_2} + \mathscr{A}_{4\omega_1} \cdot \mathscr{A}^*_{2\omega_1+\omega_2}$$
$$= T'_{4\omega_1} f'_{4\omega_1}(\theta) E(\omega_1)^6 E(\omega_2) \cos(\Delta\eta_1), \qquad (7.29)$$

and $\mathscr{A}_{2\omega_2}$ and $\mathscr{A}_{2\omega_1+\omega_2}$ interference,

$$\mathscr{P}(2\omega_2, 2\omega_1 + \omega_2) \propto \mathscr{A}^*_{2\omega_2} \cdot \mathscr{A}_{2\omega_1+\omega_2} + \mathscr{A}_{2\omega_2} \cdot \mathscr{A}^*_{2\omega_1+\omega_2}$$
$$= T''_{4\omega_1} f''_{4\omega_1}(\theta) E(\omega_1)^2 E(\omega_2)^2 \cos(\Delta\eta_2), \qquad (7.30)$$

where transition amplitude phase differences are $\Delta\eta_0 = \Delta\xi_0/2 + \Delta\phi$, $\Delta\eta_1 = \Delta\xi_1 - \Delta\phi$, and $\Delta\eta_2 = \Delta\xi_2 - \Delta\phi$, and $\Delta\xi_{0,1,2}$ are phase differences of corresponding continuum electron wave functions. The angular parts of the reduced transition matrix elements in (7.28–7.30) transform as

$$f_{4\omega_1}(\theta) = \cos^6\theta, \ f'_{4\omega_1}(\theta) = \cos^7\theta, \ f''_{4\omega_1}(\theta) = \cos^5\theta. \qquad (7.31)$$

for the parallel x polarization photoionization, and

$$f_{4\omega_1}(\theta) = \sin^6\theta, \ f'_{4\omega_1}(\theta) = \sin^7\theta, \ f''_{4\omega_1}(\theta) = \sin^5\theta. \qquad (7.32)$$

for the perpendicular y polarization photoionization. The total MPADs for the E_{e3} MATI peaks at energy $E_{e3} = 4\omega_1 - I_p$ are then,

$$\frac{d\sigma_{\parallel}(E_{e3})}{d\Omega} \propto \alpha_0 + \alpha_1 \cos^2\theta + \alpha_2 \cos^4\theta + \alpha_3 \cos^6\theta$$
$$+ \alpha_4 \cos^8\theta + \beta_0 \cos^6\theta \cos^2(\Delta\eta_0)$$
$$+ \beta_1 \cos^5\theta \cos(\Delta\eta_1) + \beta_2 \cos^7\theta \cos(\Delta\eta_2), \qquad (7.33)$$

for the parallel x polarization and

$$\frac{d\sigma_{\perp}(E_{e3})}{d\Omega} \propto \alpha_0 + \alpha_1 \sin^2\theta + \alpha_2 \sin^4\theta + \alpha_3 \sin^6\theta$$
$$+ \alpha_4 \sin^8\theta + \beta_0 \sin^6\theta \cos^2(\Delta\eta_0)$$
$$+ \beta_1 \sin^5\theta \cos(\Delta\eta_1) + \beta_2 \sin^7\theta \cos(\Delta\eta_2), \qquad (7.34)$$

for the perpendicular y polarization. From (7.33) and (7.34) we see that both *odd* and *even* powers of $\cos\theta/\sin\theta$ and $\cos(\Delta\eta)$ are obtained in the interference terms, i.e., an even number of transition $\cos^6\theta$ and $\sin^6\theta$ terms occur for odd-odd parity interference whereas the terms $\cos^5(\theta)/\sin^5(\theta)$ and $\cos^7(\theta)/\sin^7(\theta)$ correspond to odd-even transition interferences. In (7.33) and (7.34), the terms $\cos^6\theta\cos^2(\Delta\eta_0)$ and $\sin^6\theta\cos^2(\Delta\eta_0)$ come from the interference between the four ω_1 and two ω_2 photons ionization processes, which is symmetric in both angle θ and phase difference $\Delta\eta_0$. Thus the interference between $\mathscr{A}_{4\omega_1}$ and $\mathscr{A}_{2\omega_2}$ does not contribute to the asymmetry of the angular distribution. The asymmetry from the $\cos(\Delta\eta)$ $[\cos(\Delta\phi)]$ phase term only appears in the high odd multi-photon terms $\cos^5\theta/\sin^5\theta$ and $\cos^7\theta/\sin^7\theta$ via interferences of $\mathscr{A}_{2\omega_1+\omega_2}$ with $\mathscr{A}_{2\omega_2}$ and $\mathscr{A}_{4\omega_1}$ transition amplitudes, respectively.

References

1. A.H. Zewail, J. Phys. Chem. A **104**, 5660 (2000)
2. A.H. Zewail, Science **328**, 187 (2010)
3. A. Stolow, J.G. Underwood, Adv. Chem. Phys. **139**, 497 (2008)
4. P. Hockett, C.Z. Bisgaard, O.J. Clarkin, A. Stolow, Nature Phys. **7**, 612 (2011)
5. Y. Arasaki, K. Takatsuka, K. Wang, V. McKoy, J. Chem. Phys. **132**, 124307 (2011)
6. T.N. Rescigno, N. Douguet, A.E. Orel, J. Phys. B: At. Mol. Opt. Phys. **45**, 194001 (2012)
7. G.B. Griffin, S. Ithurria, D.S. Dolzhnikov, A. Linkin, D.V. Talapin, G.S. Engel, J. Chem. Phys. **138**, 014705 (2013)
8. J. Manz, L. Wöste, *Femtosecond Chemistry* (VCH, Weinheim, 1995)
9. F. Krausz, M. Ivanov, Rev. Mod. Phys. **81**, 163 (2009)
10. Z. Chang, P. Corkum, J. Opt. Soc. Am. B **27**, B9 (2010)
11. K. Zhao, Q. Zhang, M. Chini, Y. Wu, X. Wang, Z. Chang, Opt. Lett. **37**, 3891 (2012)
12. A.D. Bandrauk, S. Chelkowski, D.J. Diestler, J. Manz, K.J. Yuan, Int. J. Mass Spectrom. **277**, 189 (2008)
13. M. Vrakking, Nature **460**, 960 (2009)
14. T. Popmintchev, M.-C. Chen, P. Arpin, M.M. Murnane, H.C. Kapteyn, Nature Photon. **4**, 822 (2010)
15. R.A. Ganeev, Laser Phys. **22**, 1177 (2012)
16. B.W.J. McNeil, N.R. Thompson, Nature Photon. **4**, 814 (2010)
17. S. Chelkowski, G.L. Yudin, A.D. Bandrauk, J. Phys. B: At. Mol. Opt. Phys. **39**, S409 (2006)

7 Coherent Electron Wave Packet

18. H. Niikura, D.M. Villeneuve, P.B. Corkum, Phys. Rev. Lett. **94**, 083003 (2005)
19. J.P. Marangos, S. Baker, N. Kajumba, J.S. Robinson, J.W.G. Tisch, R. Torres, Phys. Chem. Chem. Phys. **10**, 35 (2008)
20. H.C. Shao, A.F. Starace, Phys. Rev. Lett. **105**, 263201 (2010)
21. F. Kelkensberg, A. Rouzée, W. Siu, G. Gademann, P. Johnsson, M. Lucchini, R.R. Lucchese, M.J.J. Vrakking, Phys. Rev. A **84**, 051404(R) (2011)
22. K.J. Yuan, H.Z. Lu, A.D. Bandrauk, Phys. Rev. A **80**, 061403(R) (2009)
23. K.J. Yuan, H.Z. Lu, A.D. Bandrauk, Phys. Rev. A **83**, 043418 (2011)
24. J. Mauritsson, T. Remetter, M. Swoboda, K. Klünder, A. L'Huillier, K.J. Schafer, O. Ghafur, F. Kelkensberg, W. Siu, P. Johnsson, M.J.J. Vrakking, I. Znakovskaya, T. Uphues, S. Zherebtsov, M.F. Kling, F. Lépine, E. Benedetti, F. Ferrari, G. Sansone, M. Nisoli, Phys. Rev. Lett. **105**, 053001 (2010)
25. P. Ranitovic, X.M. Tong, C.W. Hogle, X. Zhou, Y. Liu, N. Toshima, M.M. Murnane, H.C. Kapteyn, Phys. Rev. Lett. **106**, 193008 (2011)
26. N.N. Choi, T.F. Jiang, T. Morishita, M.H. Lee, C.D. Lin, Phys. Rev. A **82**, 013409 (2010)
27. J.C. Baggesen, L.B. Madsen, Phys. Rev. A. **83**, 021403(R) (2011)
28. K.J. Yuan, A.D. Bandrauk, Phys. Rev. A. **85**, 013413 (2012)
29. J. Feist, S. Nagele, R. Pazourek, E. Persson, B.I. Schneider, L.A. Collins, J. Burgdörfer, Phys. Rev. Lett. **103**, 063002 (2009)
30. L.R. Moore, M.A. Lysaght, J.S. Parker, H.W. van der Hart, K.T. Taylor, Phys. Rev. A **84**, 061404(R) (2011)
31. G. Sansone, T. Pfeifer, K. Simeonidis, A.I. Kuleff, ChemPhysChem **13**, 661 (2012)
32. T. Bredtmann, S. Chelkowski, A.D. Bandrauk, Phys. Rev. A **84**, 021401 (2011)
33. S. Chelkowski, T. Bredtmann, A.D. Bandrauk, Phys. Rev. A **85**, 033404 (2012)
34. T. Bredtmann, S. Chelkowski, A.D. Bandrauk, J. Phys. Chem. A **116**, 11398 (2012)
35. A.D. Bandrauk, S. Chelkowski, T. Bredtmann, *Progress in Ultrafast Intense Laser Science IX* (Springer, Berlin, 2013) chapter 3, p. 31
36. P. Tzallas, E. Skantzakis, L.A.A. Nikolopoulos, G.D. Tsakiris, D. Charalambidis, Nature Phys. **7**, 781 (2011)
37. Y.H. Jiang, A. Rudenko, O. Herrwerth, L. Foucar, M. Kurka, K.U. Kühnel, M. Lezius, M.F. Kling, J. van Tilborg, A. Belkacem, K. Ueda, S. Düsterer, R. Treusch, C.D. Schröter, R. Moshammer, J. Ullrich, Phys. Rev, Lett. **105**, 263002 (2010)
38. P.B. Corkum, Phys. Rev. Lett. **71**, 1994 (1993)
39. I. Barth, J. Manz, Y. Shigeta, K. Yagi, J. Am. Chem. Soc. **128**, 7043 (2006)
40. I. Barth, J. Manz, Phys. Rev. A **75**, 012510 (2007)
41. F. Mauger, C. Chandre, T. Uzer, Phys. Rev. Lett. **104**, 043005 (2010)
42. F. Mauger, C. Chandre, T. Uzer, Phys. Rev. Lett. **105**, 083002 (2010)
43. M. Odenweller, N. Takemoto, A. Vredenborg, K. Cole, K. Pahl, J. Titze, L.P.H. Schmidt, T. Jahnke, R. Dörner, A. Becker, Phys. Rev. Lett. **107**, 143004 (2011)
44. N. Takemoto, A. Becker, Phys. Rev. Lett. **105**, 203004 (2010)
45. F. He, A. Becker, U. Thumm, Phys. Rev. Lett. **101**, 213002 (2008)
46. T. Zuo, A.D. Bandrauk, Phys. Rev. A **52**, R2511 (1995)
47. S. Chelkowski, A.D. Bandrauk, J. Phys. B: At. Mol. Opt. Phys. **28**, L723 (1995)
48. T. Seideman, M.Y. Ivanov, P.B. Corkum, Phys. Rev. Lett. **75**, 2819 (1995)
49. K.J. Yuan, A.D. Bandrauk, Phys. Rev. A **84**, 013426 (2011)
50. C. Huang, Z. Li, Y. Zhou, Q. Tang. Q. Liao, P. Lu, Opt. Expree **20**, 11700 (2012)
51. M. Spanner, S. Gräfe, S. Chelkowski, D. Pavičić, M. Meckel, D. Zeidler, A.B. Bardon, B. Ulrich, A.D. Bandrauk, D.M. Villeneuve, R. Dörner, P.B. Corkum, A. Staudte, J. Phys. B: At. Mol. Opt. Phys. **45**, 194011 (2012)
52. D. Akoury, K. Kreidi, T. Jahnke, Th Weber, A. Staudte, M. Schöffler, N. Neumann, J. Titze, LPhH Schmidt, A. Czasch, O. Jagutzki, R.A.C. Fraga, R.E. Grisenti, R.D. Muiño, N.A. Cherepkov, S.K. Semenov, P. Ranitovic, C.L. Cocke, T. Osipov, H. Adaniya, J.C. Thompson, M.H. Prior, A. Belkacem, A.L. Landers, H. Schmidt-Böcking, R. Dörner, Science **318**, 949 (2007)

53. H.K. Kreidi, D. Akoury, T. Jahnke1, Th. Weber, A. Staudte, M. Schöffler, N. Neumann, J. Titze, L. Ph. H. Schmidt, A. Czasch, O. Jagutzki1, R.A.C. Fraga, R.E. Grisenti, M. Smolarski, P. Ranitovic, C.L. Cocke, T. Osipov, H. Adaniya, J.C. Thompson, M.H. Prior, A. Belkacem, A.L. Landers, H. Schmidt-Böcking, R.Dörner, Phys. Rev. Lett. **100**, 133005 (2008)
54. K.J. Yuan, S. Chelkowski, A.D. Bandrauk, J. Chem. Phys. **138**, 134316 (2013)
55. K.J. Yuan, A.D. Bandrauk, Phys. Rev. A **85**, 053419 (2012)
56. K.J. Yuan, A.D. Bandrauk, J. Mod. Opt. **60**, 1492 (2013)
57. K.F. Mak, K. He, J. Shan, T.F. Heinz, Nat. Nanotech. **7**, 494 (2012)
58. A.D. Bandrauk, S. Chelkowski, Phys. Rev. Lett. **87**, 273004 (2001)
59. S. Chelkowski, A.D. Bandrauk, Appl. Phys. B **74**, S113 (2002)
60. K.J. Yuan, A.D. Bandrauk, Phys. Rev. A **84**, 023410 (2011)
61. K.J. Yuan, A.D. Bandrauk, J. Phys. B: At. Mol. Opt. Phys. **45**, 074001 (2012)
62. K.J. Yuan, A.D. Bandrauk, Phys. Rev. Lett. **110**, 023003 (2013)
63. D.S. Tchitchekova, H.Z. Lu, S. Chelkowski, A.D. Bandrauk, J. Phys. B: At. Mol. Opt. Phys. **44**, 065601 (2011)
64. T. Brabec, F. Krausz, Rev. Mod. Phys. **72**, 545 (2000)
65. A.D. Bandrauk, H.Z. Lu, in *High-Dimensional Partial Differential Equations in Science and Engineering*, ed. by A.D. Bandrauk, M. Delfour, C. LeBris, C.R.M. Lecture Series, Vol. 41 (American Mathematical Society, Philadelphia, 2007), p. 1–15
66. A.D. Bandrauk, H. Shen, J. Chem. Phys. **99**, 1185 (1993)
67. H.D. Cohen, U. Fano, Phys. Rev. **150**, 30 (1966)
68. T. Zuo, A.D. Bandrauk, P.B. Corkum, Chem. Phys. Lett. **259**, 313 (1996)
69. M.J.J. Vrakking, T.E. Elsaesser, Nature Photon. **6**, 645 (2012)
70. A.S. Baltenkov, U. Becker, S.T. Manson, A. Msezane, J. Phys. B: At. Mol. Opt. Phys. **45**, 035202 (2012)
71. K.J. Yuan, A.D. Bandrauk, J. Phys. B: At. Mol. Opt. Phys. **45**, 105601 (2012)
72. L.Y. Peng, A.F. Starace, Phys. Rev. A **76**, 043401 (2007)
73. L.Y. Peng, E.A. Pronin, A.F. Starace, New J. Phys. **10**, 025030 (2008)
74. A.D. Bandrauk, J. Ruel, Phys. Rev. A **59**, 2153 (1999)
75. Y.Y. Yin, C. Chen, D.S. Elliott, A.V. Smith, Phys. Rev. Lett. **69**, 2353 (1992)
76. Y.Y. Yin, D.S. Elliott, R. Shehadeh, E.R. Grant, E.R. Chem, Phys. Lett. **241**, 591 (1995)
77. A.D. Bandrauk, S. Chelkowski, Phys. Rev. Lett. **84**, 3562 (2000)
78. Z.M. Wang, D.S. Elliott, Phys. Rev. Lett. **87**, 173001 (2001)
79. Z.M. Wang, D.S. Elliott, Phys. Rev. A **62**, 053404 (2000)
80. K.F. Lee, D.M. Villeneuve, P.B. Corkum, A. Stolow, J.G. Underwood, Phys. Rev. Lett. **97**, 173001 (2006)
81. G.L. Kamta, A.D. Bandrauk, Phys. Rev. A **71**, 053407 (2005)
82. H.L. Kim, R. Bersohn, J. Chem. Phys. **107**, 4546 (1997)

Chapter 8
Phase Evolution and THz Emission from a Femtosecond Laser Filament in Air

Peng Liu, Ruxin Li and Zhizhan Xu

Abstract The recent progress in the study of waveform-controlled terahertz (THz) generation from air plasma at SIOM is reviewed. Carrier envelope phase (CEP) stabilized Infrared (center wavelength $\sim 1.8\,\mu m$) few-cycle laser pulses are produced through a home-built three-stage optical parametric amplifier (OPA) system pumped by a Ti:Sapphire laser amplifier and a hollow-fiber compressor filled with argon gas. By focusing the few-cycle pulses into the ambient air, THz radiation is generated from the produced filament. THz waveform can be controlled by varying the filament length and the CEP of driving laser pulses. Calculations using the photocurrent model and including the propagation effects well reproduce the experimental results, and the origins of various phase shifts in the filament are elucidated. Such waveform-controlled THz emission is of great importance due to its potential application in THz sensing and coherent control of quantum systems. For the application of measuring the CEP of few-cycle laser pulses, the evolution of THz waveform generated in air plasma provides a sensitive probe to the variation of the initial CEP of propagating intense few-cycle pulses. The number and positions of the inversion of THz waveform are dependent on the initial CEP, which is near 0.5π constantly under varied input pulse energies when two inversions of THz waveform in air plasma become one. This provides a method of measuring the initial CEP in an accuracy that is only limited by the stability of the driving few-cycle pulses.

P. Liu (✉) · R. Li · Z. Xu
State Key Laboratory of High Field Laser Physics, Shanghai Institute of Optics and Fine Mechanics, Chinese Academy of Sciences, Shanghai 201800, China
e-mail: peng@siom.ac.cn

R. Li
e-mail: ruxinli@mail.shcnc.ac.cn

Z. Xu
e-mail: zzxu@mail.shcnc.ac.cn

K. Yamanouchi et al. (eds.), *Progress in Ultrafast Intense Laser Science XI*,
Springer Series in Chemical Physics 109, DOI: 10.1007/978-3-319-06731-5_8,
© Springer International Publishing Switzerland 2015

8.1 Introduction

Air-plasma based terahertz (THz) wave generation attracts much attention as it provides unique tools for nonlinear spectroscopy, imaging and remote sensing [1–3]. As an intense femtosecond laser pulse undergoes filamentation in ambient air, axial polarized THz radiation in forward direction was generated due to the transition-Cherenkov-type radiation [4]. More intense THz radiation field (up to MV/cm range) can be produced by using two-color laser field, where four wave mixing model and photocurrent model are used to explain the underlying physics [5–9]. The spatiotemporal dynamics of THz emission from the air plasma generated by the two-color laser field has recently been discussed and the scheme of tailoring the THz emission spectrum was proposed by adjusting the tunneling ionization events [10, 11].

Alternatively, intense THz radiation can be generated from air plasma driven by intense few-cycle laser pulses. An ultra-short laser pulse that contains only a few cycles of carrier wave can be written as $E = E_0(t)\cos(\omega t + \phi)$, where $E_0(t)$ is the pulse envelope and ω is the frequency of the carrier wave. The carrier envelope phase (CEP), ϕ, which is the phase of the carrier oscillations at the instant of maximum amplitude of the pulse envelope, becomes a significant parameter in modifying the electric field of laser pulses and outcomes of the interaction with a medium [12–16]. Intense CEP stabilized few-cycle pulses have been successfully applied in the generation of isolated attosecond pulses and the steering of molecular dissociative ionization dynamics [17–21]. The THz emission from plasma by few-cycle pulses has been interpreted as a result of asymmetric transient currents by ionization induced quasi-dc current produced by the few-cycle laser pulse, as confirmed in subsequent theoretical investigations [22, 23]. By measuring the amplitude and polarity of generated THz emission, the CEP of few-cycle laser pulses can be inferred [24].

Recent investigations show that waveform-controlled THz radiation is of great importance due to its potential application in THz imaging and coherent control of molecular dynamics [25, 26]. Can we control the waveform of femtosecond laser filament based THz radiation? Understanding the propagation effects on the THz radiation from the few-cycle laser pulse produced filament is crucial for the THz generation based CEP metrology [24] and the generation of intense and waveform-controlled THz radiation. It is well understood that a converging light wave experiences a phase shift of π, the so-called Gouy phase shift through the laser focus, which has been measured using the stereo above-threshold ionization (ATI) method in high vacuum condition [27]. However, the existence of Gouy phase shift in femtosecond laser filaments in air remains controversial [28, 29]. The dispersion of the air and the induced plasma may also affect the pulse front of the propagating few-cycle pulses.

In this chapter, a three-stage optical parametric amplifier (OPA) is built to generate IR femtosecond laser pulses with passive stabilized CEP for the idler output. The laser pulses are then compressed down to 8.6 fs ($\sim$1.5 cycles at the center wavelength of 1.8 μm) by a hollow-core fiber filled with argon gas. We then propose and demonstrate a novel scheme to generate waveform-controlled THz radiation from the air filament produced by the few-cycle laser pulses. Variation of THz waveform and even its polarity inversion are found in the spatially-resolved measurement of THz emission along

8 Phase Evolution and THz Emission

the filament. THz waveform can be controlled by varying the filament length and the initial CEP of the driving laser pulses. Finally, the variation of THz waveform in air plasma can be applied to determine the initial CEP of few-cycle pulses accurately.

8.2 Generation of CEP Stabilized IR Few-Cycle Laser Pulses

In order to obtain the CEP stabilized pulses, the difference-frequency generation (DFG) technique is relatively insensitive to the environmental disturbance comparing with the active CEP stabilization technique [30, 31]. As the phase drift is naturally eliminated in the DFG process, excellent long-term CEP stability can be expected.

There are two methods for CEP stabilized pulse generation by the DFG technique: (i) the CEP stabilized seed is generated in the DFG process within a single pulse, then amplified in the following optical parametric amplifier or optical parametric chirped-pulse amplifier (OPCPA) stages. 1.2 mJ, 17 fs at 1.5 μm [32] and 740 μJ, 15.6 fs at 2.1 μm [33] CEP stabilized pulses were generated by using this method; (ii) the CEP stabilized idler pulse is generated in an OPA process directly, if the pump and the signal pulses are from the same laser source with the same carrier phase offset. 1.5 mJ, 19.8 fs at 1.5 μm [34] and 0.4 mJ, 11.5 fs at 1.8 μm [35, 36] CEP stabilized pulses are obtained in this way. We adopt the second scheme to generate CEP stabilized few-cycle pulses by an OPA system followed by a hollow fiber compressor. Firstly, we obtained CEP stabilized 0.9 mJ, 40 fs pulses at 1.8 μm from a three-stage near-IR OPA pumped by a commercial 5.2 mJ, 25 fs Ti:sapphire laser amplifier at 1 kHz repetition rate. The obtained 1.8 μm pulses are then spectrally broadened by the nonlinear propagation in an argon-filled hollow fiber and subsequently compressed to 8.6 fs (1.5 cycles) with energy of 0.5 mJ simply using a pair of thin fused silica wedges. The compressed pulses are CEP stabilized with ~570 m rad rms CEP fluctuations, making this source a suitable driver for attosecond pulse generation. Nonlinear spectral interferometry is employed to test the CEP stability.

8.2.1 Optical Parametric Amplifier

The configuration of the three-stage OPA is shown in Fig. 8.1. The pump laser pulses are from a commercial Ti:sapphire laser amplifier (Coherent Elite-HP-USX) which provides 25 fs pulses with pulse energy up to 5.2 mJ at 1 kHz repetition rate. The pump pulses are divided into four parts using three beam splitters. The smallest part of the laser pulses (<20 μJ) is focused into a 2-mm-thick sapphire plate to generate a single-filament white light continuum (WLC), which is used as the seed pulses for the following OPA stages.

A fraction of the pump laser pulses with ~80 μJ pulse energy is used to pump the first near-collinear OPA stage (OPA1) consisting of a 2.5-mm-thick BBO crystal cut for type II phase matching ($\theta = 27.2°$, $\varphi = 30°$). The intersection angle between

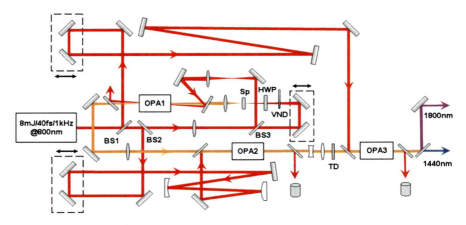

Fig. 8.1 Experimental setup for the generation of high energy self-phase-stabilized pulses: *VND*, variable neutral density filter; *HWP*, half-wavelength waveplate; *Sp*, sapphire plate; *TD*, time delay crystal [38]

the pump and seed beams is ∼1°. The WLC is amplified to ∼3 μJ in OPA1 with the center wavelength at 1.44 μm. Although OPA with type I BBO crystal in this spectral region has broader phase matching bandwidth [37], we found a lower efficiency and strong parasitic self-diffraction.

The amplified pulses at 1.44 μm from OPA1 are collimated and injected into the second collinear OPA stage (OPA2) consisting of a 2-mm-thick BBO crystal cut for type II phase matching ($\theta = 27.2$, $\varphi = 30$). About 0.7 mJ laser pulse with a diameter of ∼4 mm is used to pump OPA2. In order to get high efficiency and prevent wave front tilt [39], both the signal and the pump beams are well collimated and the two beams are collinearly injected into the BBO crystal. The seed pulses are amplified to ∼50 μJ in this stage.

The amplified signal pulses from OPA2 are enlarged and collimated to ∼9 mm in diameter with a Galilean telescope. A 1.5-mm-thick a-cut YVO$_4$ crystal is employed as a time delay crystal to separate the signal and the idler pulses in time. Then the laser beam is injected into the third collinear OPA stage (OPA3) consisting of a 2-mm-thick BBO crystal cut for type II phase matching ($\theta = 27.2$, $\varphi = 30$). The remaining ∼3.4 mJ pump laser with diameter of ∼9 mm is used to pump OPA3. Same as OPA2, the pump and the signal beams are firstly collimated and then collinearly injected into the BBO crystal. This collinear setup is quite necessary to avoid angular dispersion in the generated idler beam. The signal pulses are further amplified to ∼1.0 mJ at 1.44 μm, corresponding to the idler pulses with energy of ∼0.9 mJ at 1.8 μm. Usually, it is necessary to optimize the grating-based compressor (both the angle of incidence and the separation distance) in the Ti:sapphire laser system to optimize the chirp of the pump laser to obtain this high conversion efficiency. The pulse duration of the output signal pulses and the idler pulses are measured when the highest conversion efficiency is obtained by a home-built single-shot autocorrelator. The idler pulse duration is ∼40 fs at 1.8 μm without any further pulse compression.

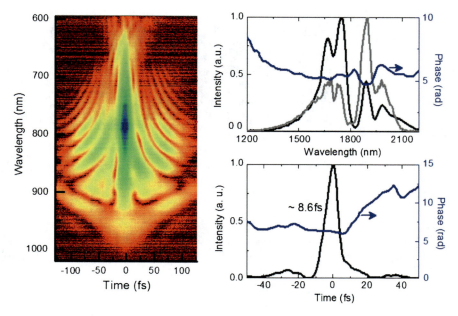

Fig. 8.2 *Left* measured SHG-FROG trace; *up right* measured (*grey*) and retrieved (*black*) spectral intensity and retrieved phase (*blue*); *down right* retrieved temporal intensity (*black*) and phase (*blue*)

8.2.2 Pulse Compression

The 1.8 μm idler beam from above OPA system is coupled into a 1-m long hollow fiber filled with argon (400 μm inner diameter, 0.5-mm-thick fused silica window) using a $f = 0.75$ m plano-convex lens. The output beam is collimated with an $R = 2$ m silver-coated concave mirror. For a balance between broad spectrum and single mode purity of the output beam, 500 mbar of argon gas is filled in the hollow fiber. In this case, supercontinuum in the range of 1,200–2,100 nm is obtained which supports the Fourier transform-limited pulse duration of about 7.8 fs. A pair of fused silica wedge is used to compensate the pulse dispersion due to its negative group delay dispersion (GDD) in this spectral range [35]. After compression, the output energy reaches 500 μJ per pulse.

Characterization of pulse duration is carried out with a home-built second harmonic generation frequency resolved optical grating (SHG-FROG), and a home-built autocorrelator. No transmissive optics elements are used in the SHG-FROG which makes it suitable to measure the pulse duration down to a few cycles. A 20-μm-thick type I BBO crystal ($\theta = 20.2$) is used in the SHG-FROG which ensures broad SHG bandwidth. In the optimized condition, the measured pulse duration is about 8.6 fs, as shown in Fig. 8.2.

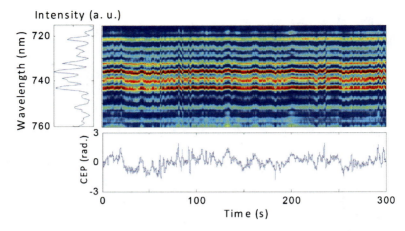

Fig. 8.3 CEP measurement: *left*, interference fringes of fundamental and second harmonic components; *up right*, sequence of interferograms acquired over 300s; *down right*, the evolution of calculated CEP

8.2.3 Carrier Envelope Phase Stability

The CEP stability is characterized with a home-built collinear $f - 2f$ interferometer whose layout is similar to the reference [40]. The only difference is a 5-mm-thick c-cut LiNbO$_3$ crystal inserted between the WLC and SHG crystal. Because the group delay is almost the same for 900 and 1800 nm lights in glass, the fringe period can not be observed if the time delay crystal is not used. Frequency overlap between the spectral broadened fundamental frequency and the second harmonic frequency is achieved in the 800–900 nm spectral range. Figure 8.3 shows the CEP fringes and phase measurement for the compressed 1.8 μm idler pulses over a 300 s observation time, whose fluctuation is ∼570 m radian.

8.3 Waveform Controlled THz Emission from Air Plasma Driven by Few-Cycle Pulses

We propose and demonstrate a novel scheme to generate waveform-controlled THz radiation from the air filament produced by few-cycle laser pulses. Variation of THz waveform and even its polarity inversion are found in the spatially-resolved measurement of THz emission along the filament. THz waveform can be controlled by varying the filament length and the initial CEP of the driving laser pulses.

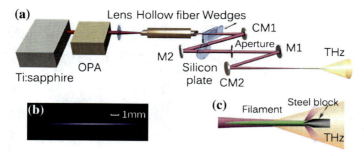

Fig. 8.4 a Schematics of the experimental layout; **b** Image of the filament formed by focusing the few-cycle laser pulse into ambient air; **c** Schematics of the filament length control [41]

8.3.1 Variation of THz Waveform in Air Plasma

In order to measure the THz time domain spectrum (TDS), the output beam of the Ti:sapphire laser amplifier is split into two beam: a major part of the beam energy, 4.6 mJ, is used to pump the OPA system while the leftover part as a probe pulse for sampling the generated THz temporal waveforms. In routine experiments, the 40 fs-long CEP-stabilized laser pulses at 1.8 µm are compressed to averaged 10 fs with the maximum pulse energy of 450 µJ. When the laser pulse is focused into ambient air by using a spherical mirror $f = 150$ mm, a stable luminescence filament of ~12 mm-long is formed by using 300 µJ pulse energy by an iris diaphragm, as shown in Fig. 8.4b. The generated THz radiation are measured by using the standard balanced diode geometry electric-optic (EO) sampling technique.

In order to investigate the evolution of the driving laser pulse and THz radiation in the filament, we insert a sharp stainless steel blade (~0.2 × 4 × 20 mm) into the plasma column to stop the filament, which can be moved by a motor stage along the laser propagation direction defined as the z coordinate henceforth. As shown in Fig. 8.4c, the blade is positioned to allow the laser beam hit onto its sharp edge, so that the detection of THz radiation from the leftover filament is less influenced. In so doing, the integrated THz signals from filaments of different lengths are measured. We calibrate experimentally the deviation of the collection efficiency of the THz detection optics varied by the filament length to the detector.

Figure 8.5a shows two THz waveforms by blocking the filament at the position of 3 mm (the distance from the visible starting position of the filament to the blade) and 10 mm, respectively. It is reasonable that the measured THz emission from the 3 mm-long filament is much weaker than from the 10 mm-long one, since the measured signal comes from the integration of the unblocked filament. One can also note that the recorded THz radiation from the 10 mm-long filament reverses its polarity comparing to that from the 3 mm-long filament, indicating a phase variation of the driving laser field in the filament. By moving the blocking blade continuously in a step length of 0.5 mm, the THz waveforms as a function of filament length are recorded and plotted in Fig. 8.5b. One can see that, the THz emission gradually changes its

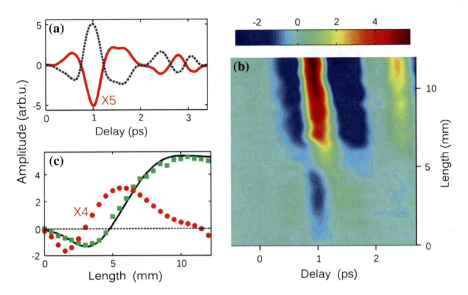

Fig. 8.5 a THz waveforms measured from the filaments of different length 3 mm (*red solid line*) and 10 mm (*black dashed line*), respectively; **b** THz waveforms measured by moving the block continuously along the filament; **c** The measured THz amplitudes (*green solid squares*), the calculated results (*black solid line*), and the retrieved THz signal amplitude (*red solid circles*) as a function of the position in the filament [41]

amplitude and polarity when the length of filament is varied, and at certain length (~5 mm-long filament) the THz emission signal goes to zero and then reverses the polarity thereafter. One can therefore control the THz waveform by varying the filament length.

For analyzing the amplitude modulation of THz radiation, the amplitude values when the THz signal is the maximum, at ~1 ps in Fig. 8.5b, are plotted in Fig. 8.5c. With the help of the above-mentioned calibration of collection efficiency, the spatially resolved THz emission as a function of the position in the filament can be retrieved, shown in Fig. 8.5c. One can see that the THz emission first increases in one (negative) polarity and then decreases to zero at ~3 mm, after which the amplitude increases in the other (positive) polarity and decreases at the tail of the filament.

8.3.2 Simulation of THz Emission in Air Plasma by a Few-Cycle Pulses

The polarization of THz emission was measured to be nearly linear so that we excludes that the generated THz emission originates from transition-Cherenkov-type radiation. We consider the THz emission originate from the transient photocurrent driven by the propagating intense few-cycle laser fields in the plasma [6, 7].

8 Phase Evolution and THz Emission

The propagation of intense few-cycle laser fields in a dispersive medium can be described by using a propagation equation in an axial-symmetric coordinates [42],

$$\partial_z \tilde{E}(r, z, \omega) = [\frac{i}{2k(\omega)}\nabla_\perp^2 + ik(\omega)]\tilde{E}(r, z, \omega) + \frac{i\omega^2}{2c^2\varepsilon_0 k(\omega)}\tilde{P}_{NL} - \frac{\omega}{2c^2\varepsilon_0 k(\omega)}\tilde{J}_{ioni},$$

(8.1)

where $\tilde{E}(r, z, \omega)$ is the frequency domain laser field, the nonlinear polarization $\tilde{P}_{NL}$ accounts for the Kerr effect and the polarization $\tilde{J}_{ioni}$ is caused by photoelectrons from the tunneling ionization of N_2 and O_2 in air. The collective motion of the tunneling ionized electrons results in a directional nonlinear photocurrent surge described by [11]

$$\partial_t J_e(r, z, t) + v_e J_e(r, z, t) = \frac{e^2}{m}\rho_e(r, z, t)E(r, z, t),$$

(8.2)

where v_e, e, m and ρ_e denote the electron-ion collision rate, electron charge, mass, and electron density, respectively. The transient current at each propagation step of the calculation is treated as radiation source of far field THz emission [43]

$$E_{THz}(r', t) = -\frac{1}{4\pi\varepsilon_0}\int\frac{1}{c^2 R}\partial_t J_e(r, z, t_r)d^3r,$$

(8.3)

where R is the distance between the THz point source and the detection plane. Equation (8.3) describes the THz radiation towards all directions. The forward far field THz emission near the propagation axis from the unblocked air plasma has been added up to simulate the measured results. The initial few-cycle laser field is of a Gaussian beam with the waist size of $w_0 = 260\,\mu m(1/e^2)$ at center wavelength of 1.8 μm, duration of $\tau = 10\,fs$ in full width at half maximum (FWHM) and pulse energy of 300 μJ. By optimizing the initial CEP of the few-cycle laser field to be $\varphi_0 = 0.33\pi$, the THz amplitude as a function of filament length is found consistent with the measured results, as shown in black solid line in Fig. 8.5c.

For further understanding the THz emission characteristics in the air plasma, we show in Fig. 8.6a the space-resolved THz emission calculated from (8.3) along the filament. As one can see that the THz amplitude modulates from negative polarity at the beginning of plasma to positive one at the end, which results in the inversed THz emission polarity shown in Fig. 8.5c. In a simplified picture, the d.c. transverse current is related to discrete electron ionization events and the pulse vector potential $A(t_i)$ modulation [11, 44], $J_e \propto \sum_i \delta\rho_i A(t_i)$. In order to distinguish electrons with opposite drifting velocities, we define two quantities J_e^+ and J_e^-, which describe the positive current density and negative current density, respectively. We calculate the asymmetry of the summed current densities $\sum(J_e^+ - J_e^-)$ which matches the THz emission as shown in Fig. 8.6a.

The laser fields, electron densities and transient photocurrents at the three positions A, B and C labeled in Fig. 8.6a are calculated and plotted in Fig. 8.6b–d, respectively. At the position of A, where the THz emission is the maximum in the negative polarity,

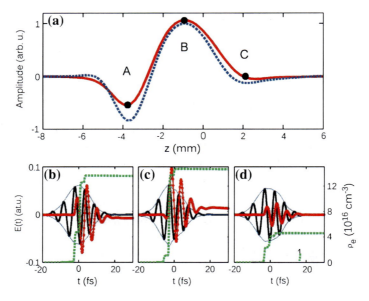

Fig. 8.6 a The calculated THz amplitude (*red solid line*) and asymmetry parameter (*blue dashed line*), $\sum (J_e^+ - J_e^-)$, along the air plasma filament; (**b–d**) At three selected positions, *A*, *B* and *C* as shown in (**a**), the laser field envelopes (*blue thin line*), carrier oscillations (*black thick line*), electron densities (*green dashed line*) and transient photocurrents (*red dotted solid line*) [41]

the amplitude of negatively current density J_e^- is larger than the positively current density J_e^+ following the laser field oscillation. At the position of B, the negative current density is less than the positive one so that the THz radiation is positively polarized. At the position of C, the almost equal current densities in the oscillation result in the near zero THz radiation. From Fig. 8.6b–d, one can also note that the increments of produced electron density ρ_e (green dashed lines) are mainly from two attosecond bursts in the opposite directions that are originated from the tunneling ionization in a single optical cycle, making the THz emission extremely sensitive to the variation of driving laser fields. The attosecond electron bursts can be easily controlled by varying the intense few-cycle fields.

8.3.3 Variation of CEP and Phase of Few-Cycle Pulses in Filament

Since the THz emission is shown to be sensitively determined by the intense few-cycle laser field, we then investigate the CEP variation in the air plasma. As indicated previously [45], the CEP is the difference of the pulse phase, i.e. carrier phase, and the pulse front. The carrier phase shift in a dispersive medium can be written as

$$\phi_{\omega_0}(\mathbf{r}, z) = n(\mathbf{r}, z)\frac{\omega_0}{c}z + \phi_G(z, z_{eff}) + \phi_{off-a}(\mathbf{r}, z) \tag{8.4}$$

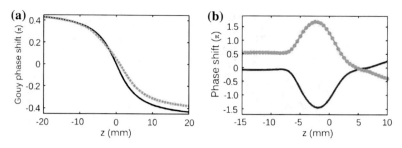

Fig. 8.7 a The modified Gouy phase shift (*red dashed line*) and the Gouy phase shift (*black solid line*) of linear focus condition; **b** The pulse front shift (*blue solid line*) and the CEP shift (*red dotted solid line*) of the laser field in air plasma [41]

where the first term is the phase shift at the carrier frequency induced by spatially dependent refractive index $n(r, z)$, the second term is the Gouy phase shift with the effective focal length z_{eff}, and the third term denotes the off-axial phase shift due to diffraction effects. For simplicity we only consider the axial phase shift that includes the first two terms. We define the second term as the modified Gouy phase shift that is characterized by z_{eff}, since the Rayleigh length is no longer valid in filament.

The calculated modified Gouy phase shift decreases with a relatively slower slope than the Gouy phase shift under the linear focus condition, as shown in Fig. 8.7a. The modified Gouy phase shift due to the extension of filament confirms the previous theoretical prediction that the Gouy phase shift stems from transverse spatial confinement of a finite beam [46]. The CEP of laser fields is also determined by the propagating pulse front, which is defined as (17) of the [45] and can be retrieved from the calculation results. The axial pulse front is shown in Fig. 8.7b. As one can see the pulse front experiences a jump around the center of the plasma region ($z = -7$ mm ~ 4 mm), as a result of the plasma-induced dispersion. The total axial CEP shift is $\sim 1.4\pi$ and possessing a hump structure in the plasma region, which is originated from the modified Gouy phase shift and plasma effects.

Furthermore, we look into the dependence of the THz emission on the initial CEP φ_0. By increasing the initial CEP gradually in a step size of 0.2π, we obtained a series of 2-dimensional THz waveform maps similar to that shown in Fig. 8.5b. The amplitudes of THz emission along the filament from experiments are retrieved and plotted as a function of φ_0 in Fig. 8.8a. We found that the THz waveforms change back to be the same as those shown in Fig. 8.5b after a change of initial CEP by 2π. This indicates that in air plasma the propagation effect on the CEP variation is independent of the initial CEP of driving laser fields, which is also verified by our numerical simulations. The phases of few-cycle pulses experience a fixed shift of larger than π, which is consistent with the CEP shift through the air plasma. This observation helps to validate the CEP metrology based on THz generation [24]. However, due to the variation of CEP through the plasma, an offset of CEP must be determined and taken into account for measuring the CEP of laser pulses. As shown in Fig. 8.8b, the simulated results using the photocurrent model and propagation

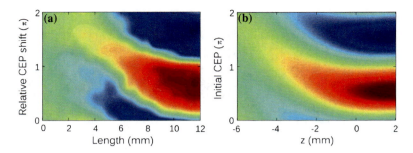

Fig. 8.8 a Measured and b simulated THz amplitude as functions of relative CEP shift and filament length [41]

equation agree well with experimental results. Since the polarity of THz radiation can be controlled by varying the initial CEP of driving laser pulses, one can maximize the output THz radiation intensity from the filament by optimizing the initial CEP to minimize the intensity cancellation due to opposite polarities.

8.4 Initial CEP and Its Determination Through THz Waveform Variation

For fully characterizing the phase of few-cycle laser pulses, several schemes of measuring the actual CEP of few-cycle pulses have been proposed and demonstrated. The stereo-ATI method measures the ATI electrons in opposite directions along the laser polarization and relates the asymmetry of cut-off electrons to the CEP value [47]. The recent advance indicates a capability of single-shot measurement with an accuracy of ∼100 milli-radian [48]. Another method is the detection of THz emission from air plasma driven by intense few-cycle laser fields, based on the CEP dependent asymmetry of tunneling ionization [24]. Thirdly, the high-harmonic emission at individual half-cycles of a few-cycle pulse has been shown to be able to measure the CEP of the driving laser field in situ [49]. These methods all aim to measure the CEP of the localized laser field where intense few-cycle pulses cross with the detection medium by assuming the variation of CEP is negligible during the interaction.

8.4.1 Initial CEP

For the intense few-cycle laser fields, Gouy phase shift and diffraction dominate the CEP variation even under the condition of linear focusing [45]. For example, if one considers few-cycle laser pulses of 8 *fs* in full width at the half maximum (FWHM)

8 Phase Evolution and THz Emission 187

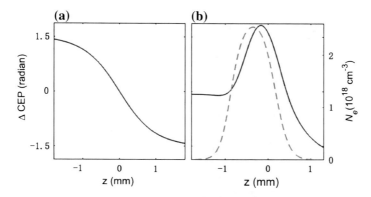

Fig. 8.9 **a** The CEP shift (*blue solid line*) around the focus of a few-cycle pulse in vacuum. **b** The CEP shift (*blue solid line*) in the interaction region between an intense few-cycle laser field and N_2 molecules, and the electron density in the air plasma (*red dashed line*) [50]

and beam diameter of 8 mm, focused by a mirror of $f = 200$ mm into vacuum, the CEP shift over a distance of 1 mm at the center of focus is 1.43 rad according to the (21) of [45], as shown in Fig. 8.9a. Interacting with a nonlinear medium, the variation of CEP becomes more complex because of the dispersion of the medium and the plasma effect. Taking the N_2 molecules (with the peak pressure of 100 torr, thickness of 1 mm) ionized by the intense few-cycle laser fields (with the pulse energy of 4 µJ and assuming the diameter of filament is 140 µm), the CEP shift over the interaction length is calculated by solving the propagation equation [42] and plotted in Fig. 8.9b. One can see that the CEP variation over the interaction length of 1 mm is 1.34 radian, and the value becomes larger as the focusing length decreases and the pulse energy increases.

An alternative option of characterizing the CEP of few-cycle laser pulses is to define the *initial* CEP, which is the actual CEP value of few-cycle pulses before the laser focusing. The initial CEP can be experimentally determined, as will be shown in this work, and its value is free of the uncertainties caused by the focusing and interaction with nonlinear medium. With the knowledge of the initial CEP, the actual CEP of the local laser field interacting with a medium can be estimated in applications.

It is known that the THz waveform reverses its polarity in air plasma because of the variation of the phase and the pulse front (namely the CEP) of the driving few-cycle fields [41]. One can in principle determine the initial CEP of driving few-cycle pulses by taking into account the variation of CEP in the air plasma. However, the value of CEP shift changes as the laser pulse energy varies, so it cannot be pre-determined in experiments and therefore hinders the accurate measurement of the initial CEP. We find that the positions where THz waveform reverses are related to the initial CEP of the intense few-cycle fields, from which a method of measuring the initial CEP of few-cycle pulses is given.

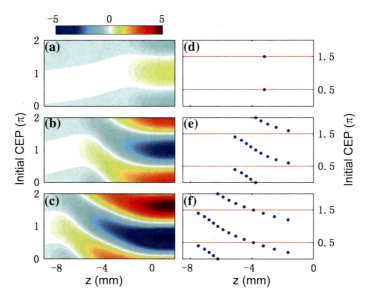

Fig. 8.10 a–c The calculated THz amplitude as a function of plasma length and the initial CEP of few-cycle pulses with the input energy of 109 µJ, 250 µJ, 400 µJ, respectively. **d–f** The inversion positions of THz emission in a step-size of 0.1 π, for the energy of 109 µJ, 250 µJ, 400 µJ, respectively [50]

8.4.2 Determination of the Initial CEP

We focus on the investigation of inversion number and positions of the THz waveform from different length of air plasma. THz emissions from air plasma can be described by the integration of far field emission from the transient photocurrents driven by an intense few-cycle laser field propagating in the plasma [6, 7, 11, 43] as in (8.3). The integrated THz amplitude as a function of filament length and initial CEP at different input energy is shown in Fig. 8.10. It can be seen that the THz amplitude, the number of the THz waveform inversion, and the inversion positions are all related to the initial CEP. As shown in Fig. 8.10a, the white color represents that the THz amplitude is zero. At a low input energy of 113 µJ, one may note that the variation of THz amplitude tilts a little when the initial CEP is near $\varphi_0 = 0.5\pi$. For the initial CEP apart from $\varphi_0 = 0.5\pi$, the THz amplitude increases monotonically along the filament. The positions of THz inversion are extracted for all the initial CEP values in Fig. 8.10d, which shows inversion of THz waveform only at 0.5π and 1.5π. It is known that because of the shift of pulse front and phase in air plasma, the variation of CEP results in varied THz waveforms [41]. Therefore the integrated far field THz emission shows a reversed polarity when the CEP of laser pulses changes across 0.5π. As shown in Fig. 8.12a, THz emission as a function of plasma length for $\varphi_0 = 0.4\pi$ and 0.6π remains positive or negative because the variation of CEP in plasma is not

8 Phase Evolution and THz Emission 189

significant enough. Therefore only the weaker THz emission from the initial CEP of $\varphi_0 = 0.5\pi$ may result in the inversion of THz polarity.

Increasing the energy of laser pulses to $250\,\mu J$, the starting position of THz generation moves to the upstream of the air plasma because of Kerr effect, as shown in Fig. 8.10b and e. Inversions of THz emission begin to appear for some initial CEP of few-cycle pulses, because of the increased CEP variation in air plasma. It shows that in the plasma the THz emission crosses with zero only once for most of the initial CEP. When the energy of laser pulses increases to $400\,\mu J$, one can see that for some initial CEP of laser pulses, $\varphi_0 = 0.2\pi - 0.4\pi$ in Fig. 8.10f, there are two inversions appear in the plasma, and the inversion positions of THz waveform gradually shift to the beginning of the filament. Finally the first inversion disappears at $\varphi_0 = 0.5\pi$. We also find that when the energy of laser pulses increases, the initial CEP of $\varphi_0 = 0.5\pi$, where the first inversion disappears into the beginning of filament, does not change.

The observation that no inversion appears in front of the plasma for the initial CEP of 0.5π indicates a smooth variation of CEP in the beginning of plasma. In the middle and end part of the filament, the significant increase and decrease of CEP results in the reversed polarity of THz emission, and consequently the integrated THz emission shows an inversion of polarity. For the intense laser fields of $\varphi_0 \geq 0.5\pi$, the THz polarity is positive at the beginning of the plasma, the superposed THz radiation keep increasing as the CEP increases in the plasma, since there is no polarity change. It eventually decreases and reverses its polarity at the downstream of the plasma when the increment of CEP is larger than $-\pi$. As a result, an inversion of THz waveform appears in the downstream of plasma. For the initial CEP of $\varphi_0 < 0.5\pi$, the THz polarity is negative at the beginning of the plasma, where the CEP can increase to be larger than 0.5π so that the instantaneous THz emission changes its polarity. As a result, the superposed THz emission becomes zero and reverses its polarity as a function of plasma length. As the energy of laser pulses increases, the variation of CEP in plasma becomes larger than π, by which the second polarity inversion of THz emission may appear if the first inversion shows at the beginning of the plasma.

Such energy-independent CEP values, $\varphi_0 = 0.5\pi$ or 1.5π, at which the inversion of THz waveforms disappears into the front of air plasma, provides a flag that one can use to determine the initial CEP of the driving few-cycle laser pulses.

8.4.3 Experimental Verification

In our experiment, the CEP stabilized infrared (IR) few-cycle laser pulses (center wavelength of $1.8\,\mu m$, 11 fs in FWHM and pulse energy of 0.46 mJ) are focused into the ambient air by a spherical mirror ($f = 150\,mm$), and a stable luminescence filament of $\sim 12\,mm$ is formed by the input pulse energy of maximum $450\,\mu J$ adjusted by an iris diaphragm. The generated THz waveforms are measured using the balanced diode geometry of electro-optical (EO) sampling method. Inserting a sharp stainless steel blade into the plasma, the THz emission from different lengths of the filament is detected. Figure 8.11a shows the THz waveforms as a function of plasma length

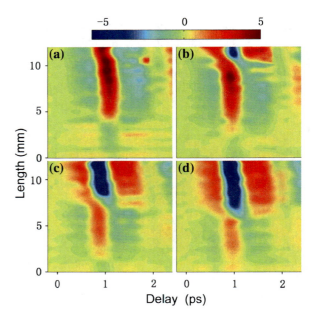

Fig. 8.11 THz waveform as a function of plasma length by few-cycle laser fields with the initial CEP of (**a–d**): $\delta\varphi_0 = 0.0\pi, 0.2\pi, 0.4\pi, 0.6\pi$, respectively [50]

when the driving few-cycle pulse energy of 400 μJ is used. The inversion of THz emission appears at the position of near 4 mm. By varying the initial CEP of the few-cycle pulses stepwise by 0.2π, the THz waveforms are recorded and plotted in Fig. 8.11b–d, respectively. It is noted that there are two inversions in the evolving THz waveform in Fig. 8.11b and c but only one in Fig. 8.11a and d. The positions of inversion are moving toward the upstream of air plasma as the initial CEP of the driving pulses increases. When the initial CEP shift increases to $\delta\varphi_0 = 0.6\pi$, the THz waveform inversion position changes back to one as the front inversion moving out of the filament, shown in Fig. 8.11d. The observation is consistent with the calculation result shown in Fig. 8.10c, f. It is reasonable to consider that the THz modulation by initial CEP of $\varphi_0 = 0.2\pi$ and 0.4π are corresponding to the conditions of the experimental observation of $\delta\varphi_0 = 0.2\pi$ and 0.4π, as shown in Fig. 8.11b and c.

The accuracy of measuring the initial CEP is determined by how well the transition from two inversions to one is decided. Even though experimentally it is limited by the CEP stability of few-cycle laser pulses, we can calculate the accurate initial CEP value where the inversion disappears at the beginning of filament. By increasing the resolution of the initial CEP to 0.001π in calculation, we find that the first inversion of THz emission disappears at the initial CEP of $\varphi_0 = 0.500\pi - 0.408\pi$ depending on the input energy (200 μJ ∼900 μJ), as shown in Fig. 8.12b. It is a slope with the fitting derivative of $-0.00013\,(\pi\,\mu J^{-1})$. This variation is due to the self-focusing effect and self-phase modulation that distorts the envelope of few-cycle pulses in plasma, leading to a little shift off the initial CEP of 0.5π.

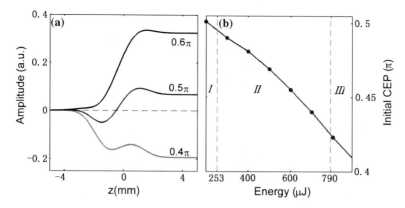

Fig. 8.12 a The calculated THz amplitude as a function of filament length for the initial CEP of $\varphi_0 = 0.4\pi, 0.5\pi, 0.6\pi$, at the input energy of 113 μJ; **b** The calculated initial CEP value by which the inversion of THz emission disappears at the beginning of air plasma under different input energies of 200 μJ ~900 μJ [50]

There are three regions, I, II and III as shown in Fig. 8.12b, for the energy of few-cycle pluses that correspond to different THz inversion phenomena. When the energy is larger than 790 μJ (III), three inversions of THz waveform can be seen for some initial CEP from our calculation result. In this case, the input energy is so high that the pulse envelope is split into two. It is therefore not realistic to discuss CEP related phenomena. For the input energy of 253 μJ–790 μJ (region II), the number of inversions of THz waveforms along the filament is one or two, corresponding to the cases in Fig. 8.10c and f. The first inversion disappears when the initial CEP is 0.427π–0.495π. For the input energy smaller than 253 μJ (region I), the number of inversions of THz waveforms along the filament is one or none for all values of the initial CEP. In this case, the disappearance of THz waveform inversion at the upstream of air plasma indicates the initial CEP is close to 0.5π. It is a good strategy to measure the initial CEP in the experimental condition shown in region I and II. In region II, it is easier to measure the THz waveform from different length of plasma since the pulse energy is higher, but the determined CEP has an error of 73 milli-radian. On the other hand, in region I, the accuracy of initial CEP can be less than 10 milli-radian if one identifies the disappearance of an inversion in the beginning of filament to be 0.5π. Therefore in practice, the measurement is only limited by the accuracy of the CEP of few-cycle laser pulses.

8.5 Summary

We have generated the CEP stabilized 8.6 fs, 0.5 mJ pulses at 1.8 μm center wavelength by a home-built OPA system followed by a hollow fiber compressor. The compressed pulses are CEP stabilized with ~570 mradian rms CEP fluctuations, making this source a suitable driver for the study of phenomena in intense laser fields. A scheme to generate waveform-controlled THz radiation from air plasma is demonstrated by using such CEP-stabilized few-cycle laser pulses. The generated THz waveform can be controlled by varying the length of the filament and the CEP of the driving laser pulses. The waveform evolution of THz radiation in the filament is due to the characteristic carrier phase shift and variation of laser envelope during propagation. Due to the ability to control the spatial location of femtosecond laser filamentation using pre-chirped laser pulses, the demonstrated scheme of waveform-controlled THz radiation generation is suitable for remote sensing exploring the phase property of THz radiation. The knowledge about the phase shift in a filament would also be valuable for optimizing the generation of phase-matched attosecond pulses driven by intense few-cycle laser pulses. We further investigate the inversion of THz emission in the air plasma driven by the intense few-cycle laser fields. The variation of the number and positions of the inversions are found dependent on the initial CEP of laser pulses. The calculation based on the transient photocurrent model indicates the first inversion vanishing point is 0.5π in the accuracy that is only limited by the CEP stability of few-cycle laser pulses. For a range of input energies of the driving pulses, the initial CEP by which the inversion disappearing at the beginning of air plasma is little energy-dependent and can be used to determine the initial CEP accurately.

Acknowledgments We acknowledge the support from National Natural Science Foundation of China (Grant Nos. 11127901, 61221064, 60978012, 11274326 and 11134010), the 973 Program of China (2011CB808103), and the State Key Laboratory of High Field Laser Physics of China.

References

1. T. Kampfrath et al., Nat. Photonics **5**, 31 (2011)
2. R. Ulbricht et al., Rev. Mod. Phys. **83**, 543 (2011)
3. J. Liu, J. Dai, S.L. Chin, X.-C. Zhang, Nat. Photonics **4**, 627 (2010)
4. C. D'Amico et al., Phys. Rev. Lett. **98**, 235002 (2007)
5. X. Xie, J. Dai, X.-C. Zhang, Phys. Rev. Lett. **96**, 075005 (2006)
6. K. Kim, J. Glownia, A. Taylor, G. Rodriguez, Opt. Express **15**, 4577 (2007)
7. K. Kim, A. Taylor, J. Glownia, G. Rodriguez, Nat. Photonics **2**, 605 (2008)
8. M.D. Thomson, M. Kreß, T.L. Loffler, H.G. Roskos, Laser Photonics Rev. **1**, 349 (2007)
9. F. Théberge et al., Phys. Rev. A **81**, 033821 (2010)
10. I. Babushkin et al., New J. Phys. **13**, 123029 (2011)
11. I. Babushkin et al., Phys. Rev. Lett. **105**, 053903 (2010)
12. A. Baltuska et al., Nature **421**, 611 (2003)
13. A. Apolonski et al., Phys. Rev. Lett. **92**, 073902 (2004)
14. T.M. Fortier et al., Phys. Rev. Lett. **92**, 147403 (2004)
15. T. Nakajima, S. Watanabe, Phys. Rev. Lett. **96**, 213001 (2006)
16. Y. Wu, X.X. Yang, Phys. Rev. A **76**, 013832 (2007)
17. M. Hentschel et al., Nature **414**, 509 (2001)
18. M.F. Krausz, M. Ivanov, Rev. Mod. Phys. **81**, 163 (2009)
19. F. Ferrari et al., Nat. Photonics **4**, 875 (2010)
20. V. Roudnev, B.D. Esry, I. Ben-Itzhak, Phys. Rev. Lett. **93**, 163601 (2004)
21. M.F. Kling et al., Science **312**, 246 (2006)

8 Phase Evolution and THz Emission

22. A.A. Silaev, N.V. Vvedenskii, Phys. Rev. Lett. **102**, 115005 (2009)
23. H.-C. Wu, J. Meyer-ter-Vehn, Z.-M. Sheng, New J. Phys. **10**, 043001 (2008)
24. M. Kreβ et al., Nat. Phys. **2**, 327 (2006)
25. W.L. Chan, J. Deibel, D.M. Mittleman, Rep. Prog. Phys. **70**, 1325 (2007)
26. K. Kitano, N. Ishii, J. Itatani, Phys. Rev. A **84**, 053408 (2011)
27. F. Lindner et al., Phys. Rev. Lett. **92**, 113001 (2004)
28. S.L. Chin, *Femtosecond Laser Filamentation* (Springer, New York, 2009)
29. Y. Liu et al., Opt. Commun. **284**, 4706 (2011)
30. A. Baltuška, T. Fuji, T. Kobayashi, Phys. Rev. Lett. **88**, 133901 (2002)
31. A. Baltuška et al., Nature **421**, 611 (2003)
32. C. Vozzi, C. Manzoni et al., J. Opt. Soc. Am. B **25**, B112 (2008)
33. X. Gu, G. Marcus et al., Opt. Express **17**, 62 (2009)
34. O.D. Mücke, S. Ališauskas et al., Opt. Lett. **34**, 2498 (2009)
35. B.E. Schmidt, P. Béjot et al., Appl. Phys. Lett. **96**, 121109 (2010)
36. P. Béjot, B.E. Schmidt et al., Phy. Rev. A **81**, 063828 (2010)
37. C. Vozzi, F. Calegari et al., Opt. Lett. **32**, 2957 (2007)
38. C. Li, D. Wang et al., Opt. Express **19**, 6783 (2011)
39. T. Kobayashi, A. Baltuška, Meas. Sci. Technol. **13**, 1671–1682 (2002)
40. A. Baltuška, M. Uiberacker et al., IEEE J. Sel. Topics Quantum Electron. **9**, 972 (2003)
41. Y. Bai et al., Phys. Rev. Lett. **108**, 255004 (2012)
42. J.S. Liu, R.X. Li, Z.Z. Xu, Phys. Rev. A **74**, 043801 (2006)
43. C. Köhler et al., Opt. Lett. **36**, 3166 (2011)
44. M. Geissler et al., Phys. Rev. Lett. **83**, 2930 (1999)
45. M.A. Porras, Phys. Rev. E **65**, 026606 (2002)
46. S. Feng, H.G. Winful, Opt. Lett. **26**, 485 (2001)
47. G.G. Paulus et al., Phys. Rev. Lett. **91**, 253004 (2003)
48. T. Wittmann et al., Nat. Phys. **5**, 357 (2009)
49. C.A. Haworth et al., Nat. Phys. **3**, 52 (2007)
50. R. Xu et al., Appl. Phys. Lett. **103**, 061111 (2013)

Chapter 9
Interaction of Femtosecond-Laser-Induced Filament Plasma with External Electric Field for the Application to Electric Field Measurement

Takashi Fujii, Kiyohiro Sugiyama, Alexei Zhidkov, Megumu Miki, Eiki Hotta and Koshichi Nemoto

Abstract The global monitoring of electrical processes in the atmosphere may become a very important part of environment control and prevention of disasters. Here we overview the theoretical and experimental studies on the radiation of laser filament plasma in an external electric field that clearly demonstrates the possibility of remote electric-field measurements. The recombination of air plasma in an external electric field depends on the field strength. The corresponding radiation of excited molecules (N_2^*, O_3^*, and others) in the ultraviolet range is also sensitive to the field strength. In spite of plasma self-radiation that occurs even without an electric field, the fluorescence grows up even at low field strengths. Air plasma of laser filaments, induced by femtosecond laser pulses, is a good candidate for the remote field measurement. The most sensitive line for the field measurement may be the 337.1 nm line representing the $N_2^*(C^3) \rightarrow N_2(B^3)$ transition. The fluorescence power at 337.1 nm is shown to increase exponentially with the field strength.

T. Fujii (✉) · A. Zhidkov · M. Miki · K. Nemoto
Electric Power Engineering Research Laboratory, Central Research Institute of Electric Power Industry, 2-6-1 Nagasaka, Yokosuka-shi, Kanagawa 240-0196, Japan
e-mail: fujii@criepi.denken.or.jp

A. Zhidkov
e-mail: zhidkov@ppc.osaka-u.ac.jp

M. Miki
e-mail: megu@criepi.denken.or.jp

K. Nemoto
e-mail: nemoto@criepi.denken.or.jp

T. Fujii · K. Sugiyama · E. Hotta
Interdisciplinary Graduate School of Science and Engineering, Tokyo Institute of Technology, 4259 Nagatsuta-cho, Midori-ku, Yokohama-shi, Kanagawa 226-8502, Japan
e-mail: ksugi@cs.trdi.mod.go.jp

E. Hotta
e-mail: hotta.e.aa@m.titech.ac.jp

K. Yamanouchi et al. (eds.), *Progress in Ultrafast Intense Laser Science XI*,
Springer Series in Chemical Physics 109, DOI: 10.1007/978-3-319-06731-5_9,
© Springer International Publishing Switzerland 2015

9.1 Introduction

The laser filament plasma (LFP) [1–23], produced by intense ultra-short laser pulses, in a strong external electric field has attracted interest both in the physics of streamer discharges [24, 25] and in various applications such as the discharge triggering [26–38], the generation of terahertz radiation [39], and the remote measurement of electric fields in air [38, 40–44].

Recently, several groups have already reported on experiments with laser-filament plasma in an external field. However, they have mostly focused on the femtosecond (fs)-pulse-laser-triggered discharges [26–38] trying to reduce the breakdown voltage, which has been successfully done. However, the dynamics of LFP before the breakdown is also surely important to understand filament physics and for applications. For example, LFP may find various applications as a point source of radiation in atmosphere: a source that is very sensitive to atmospheric conditions including atmospheric electric field. An interesting phenomenon has been found in pioneering experiment [39] on the terahertz radiation of LFP: the intensity of the radiation is very sensitive to the external field strength. The UV radiation of LFP has been also found very sensitive to external field strength [38, 40–44]: it increases nonlinearly with the applied voltage. The field dependency of UV emission may become a basis for the remote field measurements in the atmosphere and at industrial objects.

The study of atmospheric electricity may become much more advanced upon achieving the nondestructive, remote, and time-resolved measurement of the electric-field distribution. Starting with the pioneering work of Franklin [45], numerous methods and techniques have been proposed to reveal the structure and dynamics of atmospheric electricity. In addition to purely scientific problems concerning atmospheric electricity, including the global simulation activity [46], the industrial research on the active protection [37, 47, 48] of sensitive facilities from lightning is also still in its infancy. The measurement of atmospheric electricity must be extremely remote and nondestructive for a precise dynamical monitoring. However, conventional methods of detecting atmospheric electricity and its dynamics [49] are not comprehensive because of their effects on the field distribution, and usually very local to carry out atmospheric tomography. The active lightning prevention [37, 47, 48] has to detect cloud-cloud discharges; otherwise, the conversion of cloud-cloud to cloud-ground discharges makes such a method even hazardous. The detection requires a value of the instantaneous strength of the electric field before the occurrence of an electric spark.

Spectroscopy is the most efficient nondestructive technique used for studying natural phenomena. Up to now, modern laser methods such as lidar [50], being a powerful technique for atmospheric monitoring, cannot be used for electric field measurement. To detect the electric field, plasma is required. Air plasma in an electric field radiates in the ultraviolet (UV) region. The stronger the electric field, the higher the power of UV radiation. In the absence of electric discharge, seeding electrons are necessary for the plasma to emit UV light. There are various kinetic processes in air involving seeding electrons: ionization, excitation, recombination,

9 Interaction of Femtosecond-Laser-Induced Filament Plasma 197

and attachment [25]. The electric field increases the ionization and excitation rates, whereas the recombination rate decreases. Theoretically, the energy of UV emission and its duration can be used to obtain an exact value of the electric field in the space occupied by bounded plasma. The key issues are how to make a suitable plasma source of UV emission that is sensitive to the electric field strength and how to calibrate the emission versus the field strength. Laser-induced plasma appears to be a good candidate for the source. Presently, there are two types of plasma that can emit powerful UV radiation: laser-breakdown plasma (LBP) and LFP. LBP is a well-known phenomenon while LFP is a recently discovered phenomenon that has been extensively studied over the last decade.

In this chapter, we overview the kinetics and dynamics of the LFP in external electric fields, and focus on the dynamics of UV emission from LFP for the applications of remote field measurement.

9.2 Filamentation Induced by High-Intensity Femtosecond Laser Pulses and Its Interaction with External Electric Field

LFP is produced in air by fs laser pulses via laser field ionization (multiphoton or/and tunnel ionization) [51]. A fs laser pulse undergoes Kerr self-focusing in air [2, 7–9]; its intensity increases until LFP is produced. Diffraction of the laser pulse in the plasma prevents any further self-focusing. The parameters of LFP such as the electron density and temperature strongly depend on the energy of laser pulses and focusing conditions. The electron density in LFP is about 10^{16}–10^{18} cm^{-3} ([1–3, 15, 22]), and the electron temperature is low; it has been estimated to be 0.5 eV [22]. A comprehensive analysis of the laser-filament dynamics in air can be found in [17]. Recently, the emission of laser filaments at several kilometer altitudes has been detected [6], and also ionization channels have been observed over a distance reaching 400 m [13]. These results show that the remote measurement using the filament plasma is possible [31]. In addition, Kasparian et al. observed the increase of the cloud-cloud discharges using laser filament, which shows the presence of plasma at several kilometer altitude [37]. Measurability of LFP radiation, caused by breakdown of aqueous aerosol, in a kilometer distance has been shown by Daigle et al. [52].

The LFP is not a conventional discharge plasma or laser-breakdown plasma. Even in a strong electric field, LFP is cold and the dissociation recombination reduces the LFP electron density quickly: $N_e = N_{e0}/[1 + \beta N_{e0}t]$, where $\beta \sim 2 \times 10^{-7}$ cm^3/s [25, 53] and N_{e0} is the initial electron density in the filament. At $t \leq (\beta N_{e0})^{-1} \sim (5$–500 ps), the LFP is dense, the Debye radius is smaller than a plasma size; electrons do not attach to oxygen molecules. At $t > (\beta N_{e0})^{-1}$, the electron density loses the information on the initial value. For tens of nanosecond, the electron density may exceed a 'critical' electron density, $\sim 5 \times 10^{14}$ cm^{-3} [36], at which the electron attachment to molecules becomes dominant and the density of negative ions rapidly grows. In this range, discharge between the electrode and LFP, electrons avalanches and streamers heat up plasma and change filament plasma dynamics. Since the electron collision time, τ_c and drift velocity, v_D, are ~ 1 ps and $\sim 10^7$ cm/s correspondingly, plasma electrons can acquire the energy, $\varepsilon \sim m v_D^2 t / \tau_c$ [25], up to several eV and lose it in the molecule excitations provoking a powerful UV burst. Since the number of electrons able to excite the electron-electron transitions in molecules grows exponentially with the field strength, the UV emission can be very sensitive to the external field.

However, there are various effects which can also influence the LFP emission such as the secondary emission, strong corona, runaway electrons, and so on and, therefore, can change the emission dependency on the field strength. The LFP, itself, may also affect the field distribution in the vicinity of measurements. To make a practical tool, we have to prove the conformity between the emission signal and the external field strength and the ability to maintain it. Detailed studies of LFP behavior in the external fields of different configurations are necessary to understand whether such measurement can be calibrated or can provide only the relative field measurements.

9.3 Remote Measurement of Electric Field Using Filament Plasma

9.3.1 Theory

To estimate the behavior of UV emission from the LFP in an external field we performed 1D3V particle-in-cell (PIC) simulation including elastic collisions e-M, e-e (M means molecules) and the kinetic approach following [54] with variable particle weight. The electron–electron collisions were included because they are dominant in at least the first 1 ns where the electron density exceeds 10^{14} cm^{-3}. Only five sorts of particles have been included: electrons, negative oxygen ions, positive oxygen and nitrogen ions, and meta-stable nitrogen molecules. Assuming the rapid association we attribute the ions to O_2^- and $O_2^+ = X^+$ and $N_2^+ = Y^+$. The weight of particles was calculated in the kinetic cell according to the balance equation [25]. For the electron part it is,

$$\frac{dN_e}{dt} = [\alpha_{ion}^X(E/N) \, N_X + \alpha_{ion}^Y(E/N) \, N_Y]N_e + \pi_P N_{Y*}^2$$
$$- [\beta_{DR}^X N_{X^+} + \beta_{DR}^Y N_{Y^+}]N_e - v_{Atach} N_X N_e \qquad (9.1)$$

where α_{ion}^M is the ionization rate of M molecule calculated with the sampling electron distribution in PIC, π_P is the rate of Penning reaction; β_{DR}^M is the dissociative recombination rate for M ions, and v_{Atach} is the electron attachment rate. In the present calculation the Penning effect was small. In the ion part, the ion-ion recombination of positive and negative ions [25] was included as well. The UV emission was

9 Interaction of Femtosecond-Laser-Induced Filament Plasma

Fig. 9.1 The dependency of UV power per electron at 337.1 nm from nitrogen molecules on the electric field strength obtained from PIC simulations performed at different temperatures of air. The initial electron density was set $N_e(t=0) = 10^{15}$ cm^{-3} [42, 43]

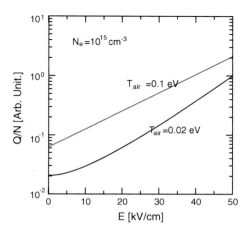

calculated for 337.1 nm transition using the simplified 2 level approximation [25] with the sampling electron distribution from the PIC.

The electric field was calculated via Poisson equation;

$$d^2\phi/dx^2 = -4\pi e (N_e - N_{X^+} - N_{Y^+} - N_{X^-}) \quad (9.2)$$

with the apparent boundary condition

$$-\frac{d\phi}{dx}(x \to \pm\infty) = E_0 \quad (9.3)$$

The result of the simulation for two different gas temperatures is illustrated by Fig. 9.1. The power Q of UV emission was normalized on the initial electron density. One can see an exponential dependency of the integrated UV power of the fluorescence on the external field strength. This dependency can be fitted as $Q/N_e \sim \exp(E/E_{\text{eff}})$ with $E_{\text{eff}} \sim 8$ and 14 kV/cm at $T_{\text{air}} = 0.02$ and 0.1 eV accordingly.

Dynamics of laser filament plasma in an external, near corona threshold field was studied by 2D3V PIC simulation using the balance equation mentioned above. We used the Buneman scheme and calculated the electric field using pair of Maxwell equations:

$$\partial\vec{E}/\partial(ct) = \vec{\nabla} \times \vec{B} - 4\pi\vec{j} \quad (9.4)$$

$$\partial\vec{B}/\partial(ct) = -\vec{\nabla} \times \vec{E} \quad (9.5)$$

with the initial conditions found from a solution of Poison equation

$$d^2\phi/dx^2 = 0; \quad -\nabla_\perp\phi|_{boundary} = \vec{E} \quad (9.6)$$

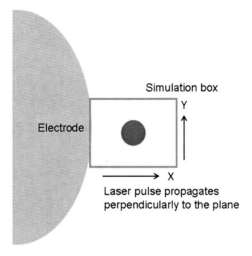

Fig. 9.2 Simulation area. The initial field is directed from the *left* to the *right*

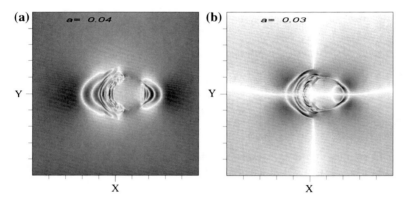

Fig. 9.3 The x-(**a**) and y-(**b**) components of electric field near the laser filament plasma of *cylindrical shape*. Diameter of the filament is 0.5 mm. $E_x(0) = 29\,\text{kV/cm}$, a is the normalized field strength; $a(0) = 0.02$ [38, 43]

The simulation geometry is shown in Fig. 9.2. Results of the simulation for three different filament plasma shapes are shown in Figs. 9.3, 9.4, and 9.5. $a = eE/mc\omega_{pl}(0)$ ($\omega_{pl}(0)$: initial plasma frequency) is the normalized field strength. One can see that electric field strength near filament plasma is strongly enhanced in the direction of external electric field.

9.3.2 Experimental Setup

A schematic diagram of the experimental setup is shown in Fig. 9.6. A Ti: sapphire laser system (Thales Laser; Alpha 10/US-20TW) was used for producing LFP.

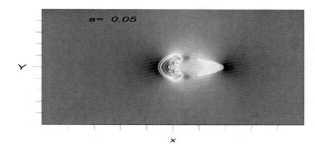

Fig. 9.4 The *x*-component of electric field near the laser filament plasma of *elliptical shape*. The filament is 0.3 × 1 mm² [38]

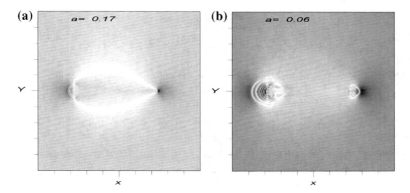

Fig. 9.5 Dynamics of *x*-component of electric field near the laser filament plasma of *needle shape*. a is the normalized field strength; a=0.17 (**a**) and a=0.06 (**b**). The filament is 0.1 × 1 mm² [38]

The Ti: sapphire laser pulses ($\lambda = 800$ nm, $\tau = 50$ fs, energy = 84 mJ) were focused by a concave mirror of 10 m focal length. A negative high voltage was applied on a spherical HVE with 60 or 250 mm diameter placed in the distance of 10.4 m from the focusing mirror. To avoid any effect of laser-induced discharge, the HVE was placed over 1 m above the grounded floor and very far from any conducting surrounding. Positive or negative dc voltage was set from 0 to ±400 kV.

The field E was non-uniform and reached its maximum on the HVE, as shown by the equation $E = -U_0 D/(2R^2)$, where U_0 is the applied voltage, D is the diameter of the spherical electrode, and R is the distance from the center of the electrode. From the equation, the field strength at the filament position under the HVE was calculated to vary from $E = 0$ kV/cm to a maximum of $E = 29.6$ kV/cm (close to corona threshold field in air) in case of using the spherical HVE with 250 mm diameter; it cannot be estimated in case of using that with 60 mm diameter because of the strong corona from HVE.

The optical axis of the laser beam was set at distances of 5 or 28 mm from the HVE. The fluorescence of LFP was collected by a telescope with a diameter of 152 mm located 20 m from the HVE and 1.3 m from the laser axis. The telescope

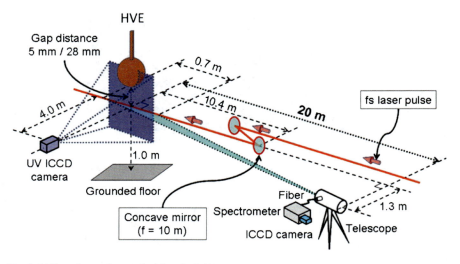

Fig. 9.6 Experimental setup for electric field measurement

measured plasma over a range of 0.7 m by centering on the HVE. The fluorescence spectra were measured by a spectrometer (Roper Scientific; SP-2358-P) and by an ICCD camera (Roper Scientific; PI-MAX: 1K-UV-MgF2). UV images of LFP were taken by another ICCD camera (Roper Scientific; PI-MAX: 1K-UniGen) with an UV lens and a filter (blocking: 400–800 nm) placed at 4 m from HVE perpendicularly to laser axis. Triggers of the both ICCD cameras were synchronized with the laser pulses. The variation of the fluorescence spectra and the UV images versus ICCD camera gate delay time from the laser shot was investigated. The delay was defined to be zero when the fluorescence signal at 337.1 nm or the UV image of filament plasma measured with the gate width of 20 ns were strongest without voltage. The spectroscopic data and the UV images were averaged over 100 laser shots for each measurement.

9.3.3 Experimental Results

9.3.3.1 Characteristics in Moderate Electric Field Using a Spherical HVE with 250 mm Diameter

We photographed transverse profiles of the laser beam with filaments; typical images are shown in Fig. 9.7. Strong Kerr self-focusing is seen even at the distance of 5 m from the concave mirror. Because of the inhomogeneous intensity profile, strong filaments are formed at the edge of the laser beam cross section. The plasma density is estimated to be $N_e = 10^{17}$ cm^{-3} near the HVE [36].

9 Interaction of Femtosecond-Laser-Induced Filament Plasma   203

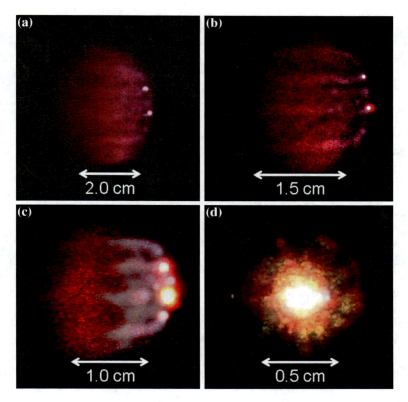

Fig. 9.7 Cross sections of fs laser pulses with filaments at distances of 5.0 m (**a**), 7.0 m (**b**), 9.0 m (**c**), and 10.7 m (**d**) from the concave mirror with 10 m focal length. The pictures were taken using *black paper* as a screen. The bright spot at the focus point shown in (**d**) is considered to consist of condensed filaments

Typical UV images of LFP in the vicinity of the HVE at different delay times of the ICCD camera gate are presented in Fig. 9.8. The optical axis of the laser was set at 28 mm from the surface of the HVE, and laser beam and filaments did not touch the HVE. Strong UV radiation can be seen in the vicinity of the filament position. One can select two zones of the radiation: the vicinity of the position at maximal electric field, in which two strong bright spots are observed, and the edges of filament plasma. The probability of the streamer formation is higher at the edges of the filament in the strongest part of the field. In the beginning, the extra charge produced at the center spreads along the filament plasma creating the edges with the streamer formation. Owing to the symmetry near the center, there are two streamers and, therefore, the two brightest spots are observed. Intensity profiles along cross section of the UV images in Fig. 9.8b and c are shown in Fig. 9.9a and b. The transverse distributions of the UV images presented in Fig. 9.9 clearly demonstrate that the most powerful signal comes from the filament. For the experimental setup in this chapter, the breakdown voltage between the HVE and the grounded floor is about 2 MV, which is much higher

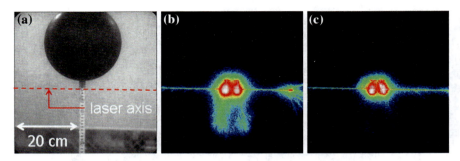

Fig. 9.8 UV images of LFP in the vicinity of HVE with 250 mm diameter when the optical axis of the laser beam was set at 28 mm from the HVE. **a** Arrangement of HVE and laser axis. **b, c** UV images in the vicinity of HVE at a voltage of −400 kV with ICCD camera gate delay time of 150 ns (**b**) and 200 ns (**c**). The gate width of the ICCD camera was 50 μs

than the maximal applied voltage used in the experiments. Therefore, there was no effect of laser-induced discharge. However, as seen in Fig. 9.8b and c, the effect of HVE exists: the junction between the HVE and the filament results in appearance of corona discharge and its consequent radiation.

The fluorescence spectra of N_2 molecules time-integrated over the gate width of 500 ns were measured for LFP as shown in Fig. 9.10a and b for different distances between filament and HVE, 5 and 28 mm. At 5 mm distance, a major increase of the fluorescence signal comparing to the self-radiation was observed at a voltage of −300 kV. The radiating part of the LFP was small because of the strong non-uniformity of the electric field in the vicinity of the laser filaments. With increasing the distance to 28 mm, the minimum detectable voltage was up-shifted. The peak signal height of the N_2 fluorescence near $\lambda = 337.1$ nm, which is the transition in the second positive band of nitrogen molecules ($N_2:C^3\Pi_u \rightarrow N_2:B^3\Pi_g$), is given in Fig. 9.10c and d for corresponding results of Fig. 9.10a and b. Again at 5 mm distance between filament and HVE, significant growth of the fluorescence signal is measured at a voltage over −200 kV. In the voltage range between −200 and −400 kV, the signal growth is nonlinear. These results show that the LFP allows the detection of very small changes in potential due to the exponential dependence. At 28 mm distance, increase of the fluorescence signal appears at a voltage of −300 kV, corresponding electric field of 16 kV/cm; its further growth is also nonlinear.

Figure 9.11 shows the peak signal height of the N_2 fluorescence over background spectra as a function of the gate delay time of the ICCD camera with the applied voltage of −400 kV for the distances between filament and HVE, 5 and 28 mm. The gate width of the ICCD camera was 50 ns. The fluorescence intensity increases when the gate delay time increases from zero to 50 ns, and decreases with the gate delay time over 50 ns. The fluorescence lasted until the gate delay time of around 350 ns. This life time of nitrogen fluorescence is much longer than that without external electric field. These results suggest that the fluorescence intensity and its lifetime can be a tool for the measurement of the electric field.

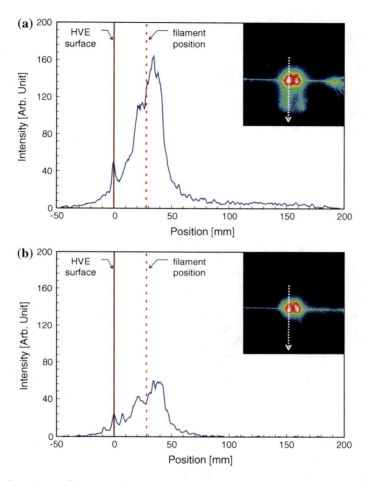

Fig. 9.9 Intensity profiles along UV image cross sections of LFP in the vicinity of HVE with 250 mm diameter when the optical axis of the laser beam was set at 28 mm from the HVE. Results as a function of the distance from HVE taken with ICCD camera gate delay time of 150 ns (**a**) and 200 ns (**b**). The applied voltage was −400 kV and the gate width of the ICCD camera was 50 μs. The cross sections were taken along *dotted arrows* as shown in *insets* of (**a**) and (**b**). The positions of the HVE surface and the filament are shown by *solid* and *dotted lines*, respectively

9.3.3.2 Characteristics in Strong Electric Field Using a Spherical HVE with 60 mm Diameter

In order to clarify the characteristics of the filament in strong corona, we have measured the emission from filament plasma using HVE with 60 mm diameter. The distance between the center of the laser beam and the HVE was 5 mm. Images of UV emission with the positively and negatively biased electrode are presented in Figs. 9.12 and 9.13, respectively. Without the laser filaments we observed no UV

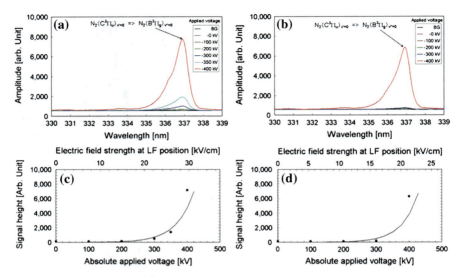

Fig. 9.10 Fluorescence spectra of N_2 molecules and peak signal height of the N_2 fluorescence (337.1 nm) for the LFP when using HVE with 250 mm diameter. **a, b** Fluorescence spectra of N_2 molecules when the optical axis of the laser beam was set at 5 mm (**a**) and 28 mm (**b**) from the HVE. These spectra were measured under different electric fields produced by the applied voltage from 0 to −400 kV. The gate width and delay time of the ICCD camera were 500 ns and 0 ns, respectively. Background spectra (shown as BG in the figures) were taken at a voltage of −400 kV without the LFP. **c, d** Peak signal height of the N_2 fluorescence (337.1 nm) over background spectra as a function of applied voltage and electric field strength at the laser filament position nearest from the HVE for the fluorescence spectra in (**a**) and (**b**) are shown in (**c**) and (**d**)

emission even at $U = +400$ kV using the ICCD camera. However, strong corona discharges were observed with the applied voltage over 200 kV using a still camera. With the laser filament the UV emission appears. In case of positively biased electrode, with a voltage increase, the intensity of UV emission in the vicinity of the HVE and LFP becomes visibly higher as seen in Fig. 9.12a. Temporal behavior of UV emission at $U = +400$ kV is shown in Fig. 9.12b. In the beginning of the laser pulse irradiation, strong UV emission appeared in the vicinity of HVE and spread about 10 cm downwards from the HVE in the first 50 ns. After 100 ns, travelling of the UV emission ceased and UV emission rapidly vanished.

Effect of the electric field on the plasma radiation can be recognized already at $U \sim 50$ kV ($E \sim -16.7$ kV/cm at the HVE surface). At the surface of HVE, a strong corona discharge may appear already at $|U| > 100$ kV, as suggested by the measurement using a still camera mentioned before, for this electrode configuration forming a negative ion layer nearby the surface. The layer thickness depends on the applied voltage. At lower voltage, $|U| < 200$ kV, it exceeds 5 mm length, therefore, the filament plasma passed through almost unscreened field.

For the negatively biased electrode, images of UV emission are given in Fig. 9.13. With the LFP, results at the voltage $|U| < 100$ kV are the same as those in case of

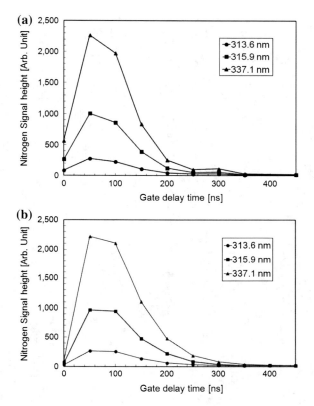

Fig. 9.11 Peak signal height of the N$_2$ fluorescence over background spectra, for the wavelengths of 313.6 nm, 315.9 nm, and 337.1 nm in the vicinity of the HVE with 250 mm diameter and filament, as a function of the gate delay time of the ICCD camera. The optical axis of the laser beam is set at 5 mm (**a**) and 28 mm (**b**) from the HVE surface for the negative polarity. The applied voltage is −400 kV and the gate width of the ICCD camera for the experiments is 50 ns [41]

the positive polarity, as shown in Fig. 9.12a. Although the region of UV emission was limited between the HVE and filament, the UV emission grows with the voltage. In Fig. 9.13b, only one image with the gate delay time of 0 ns is shown. No clear UV emission was seen with the other gate delay time after 50 ns in contrast to the spectroscopic measurement, as in Fig. 9.14a, where the UV emission was observed over 500 ns delay time in case of negative polarity. This is due to a difference in the angle of observation: the UV emission in Fig. 9.14a was integrated along the filament over a range of 0.7 m while the UV images shown in Fig. 9.13 were taken perpendicularly to laser axis. In addition, the UV emission intensity decreases rapidly as shown in Fig. 9.14a. The emission intensity at the gate delay time of 50 ns for the negative polarity is one order lower than that for positive polarity. At $U = -400$ kV, a positive ion layer screens the field with almost the same thickness dependency on the external voltage as the case with positive polarity. Here we should note that the

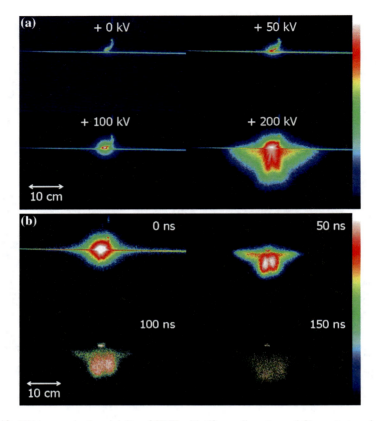

Fig. 9.12 UV images in the vicinity of HVE with 60 mm diameter and filament at each applied voltage (**a**) and each gate delay time of ICCD camera (**b**) in the case of +400 kV. The gate delay time, gate width and gain of ICCD camera for the experiments shown in (**a**) were 0 ns, 650 ns, and 200, respectively. The gate width and gain of ICCD camera for the experiments shown in (**b**) were 50 ns and 200, respectively. The intensity of UV emission was shown by *color* [40]

negative ions are repelled from the electrode creating a low density layer. However, the different layer structure makes the LFP dynamics distinct for negatively and positively biased electrodes.

In Fig. 9.14a, the temporal evolution of integrated UV emission is given for $U = \pm 400$ kV. For the positive polarity, the emission rapidly falls in 50 ns and, then, further decreases with time and almost vanishes at 300 ns. The strong emission just after the filament formation is the recombination radiation of the LFP in the external electric field. In contrast, for the negatively biased electrode, the UV emission had two distinguished stages (see Fig. 9.14a). At the first stage, the strong UV emission, occurred just after the filament formation, diminishes rapidly, much faster than that for the positive polarity, and approaches to the background level at the delay time of 100 ns. This is also the result of LFP recombination. In the second stage lasting from 100 to 500 ns, we detected a considerable growth of the emission, in order

9 Interaction of Femtosecond-Laser-Induced Filament Plasma   209

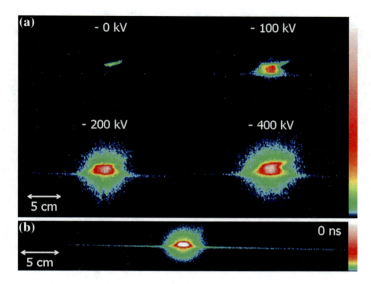

Fig. 9.13 UV images in the vicinity of HVE with 60 mm diameter and filament at each applied voltage (**a**) and at each gate delay of ICCD camera (**b**) in the case of −400 kV. The gate delay, gate width and gain of ICCD camera for the experiments shown in (**a**) were 0 ns, 650 ns, and 50, respectively. The gate width and gain of ICCD camera for the experiments shown in (**b**) were 50 ns and 100, respectively. The intensity of UV emission was shown by *color* [40, 43]

of magnitude. We attribute the signal to a discharge developing along the filament. The discharge develops after electron avalanches from the electrode provoked by the initial UV flash make a link between the positively charged LFP and the HVE, and LFP acquires the electrode potential. Consequent streamers, born in the LFP, result in its stronger ohmic heating and in the increase of its UV emission.

In Fig. 9.14b, the dependency of the UV emission power at 337.1 nm on the applied voltage is given. One can see a very rapid growth of the emission until $|U| < 200$ kV or the electric field $|E| = 66$ kV/cm at the HVE surface. Then the screening radius due to the positive or negative ions becomes too small and the LFP radiates in a smaller electric field, almost equal to the field at $|U| = 200$ kV; the UV emission from LFP saturates.

9.4 Summary

We have theoretically and experimentally studied the dynamics of filament plasma in the external electric field with a high temporal resolution. We have performed the imaging and spectral measurement of UV from LFP in the external field of different polarities produced by the spherical electrode with 25 and 6 cm diameters. The emission of excited nitrogen molecules in LFP produced in an external field is quite

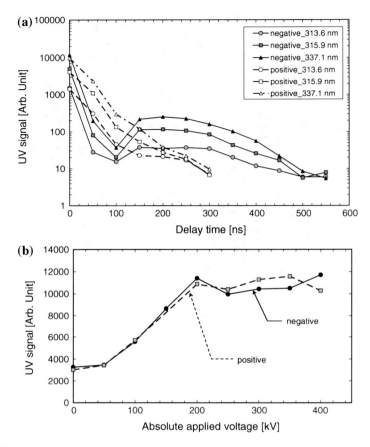

Fig. 9.14 a Time-dependencies of UV signal in the vicinity of HVE with 60 mm diameter and filament for the wavelengths of 313.6, 315.9 and 337.1 nm with the applied voltage of +400 kV (notated as positive) or −400 kV (notated as negative) and **b** dependencies of the signal height for 337.1 nm on the applied voltage for the negative and positive polarities [40]

sensitive to the strength of the field. In case of using the electrode with 25 cm diameter, the UV emission increased exponentially versus the applied voltage between −200 and −400 kV, corresponding to the electric field of 10.7 and 21.4 kV/cm at the filament position when the distance between the HVE and filament was 28 cm. In case of using the HVE with 6 cm diameter, the behavior of filament plasma at $|U| = 400$ kV with different polarities shows different dynamics; in the case of the negative polarity, the temporal evolution of UV emission supposes the discharge between the electrode and positively charged filament. The field dependency of the UV emission shows similar trend as that observed in the experiments with the large sphere electrode up to $|U| = 200$ kV or up to the electric field strength equal 66 kV/cm at the surface of the HVE. However, at the higher voltage, the strength of the UV emission was saturated due to the effect of field screening by positive or negative

9 Interaction of Femtosecond-Laser-Induced Filament Plasma

ions. The filament plasma can be created in a distance shorter than the corona screening radius. Therefore, the measurable field is higher than the corona threshold field. Since the UV emission is sensitive to the electric field, the technique presented here can become a basis for a practical field measurement; it is shown to be useful at least for the relative field measurements (detection of field distribution).

Our findings may open a very convenient and useful way for measuring the electric field and potential distributions in the atmosphere. The recombination time and the intensity of fluorescence from the LFP can be efficiently used for remote measurement. LFP has many advantages for the remote measurement of the electric field. Having a high spatial resolution, the method enables atmospheric tomography as well as the rapid detection of the field dynamics. Moreover, it was demonstrated that filaments can propagate well in adverse atmospheric conditions such as turbulence [55], foggy [56] or rainy [57] atmosphere partly due to the self-healing effect. However, precise control of the filament plasma position, which may depend on atmospheric conditions, has yet to be achieved. The necessary conformity of laboratory measurements and atmospheric measurement requires further developing.

References

1. A. Braun, G. Korn, X. Liu, D. Du, J. Squier, G. Mourou, Opt. Lett. **20**, 73–75 (1995)
2. B. La Fontaine, F. Vidal, Z. Jiang, C.Y. Chien, D. Comtois, A. Desparois, T.W. Johnston, J.-C. Kieffer, H. Pépin, Phys. Plasmas **6**, 1615–1621 (1999)
3. H. Yang, J. Zhang, Y. Li, J. Zhang, Y. Li, Z. Chen, H. Teng, Z. Wei, Z. Sheng, Phys. Rev. E **66**, 016406 (2002)
4. J. Yu, D. Mondelain, J. Kasparian, E. Salmon, S. Geffroy, C. Favre, V. Boutou, J.-P. Wolf, Appl. Opt. **42**, 7117 (2003)
5. G. Méchain, A. Couairon, M. Franco, B. Prade, A. Mysyrowicz, Phys. Rev. Lett. **93**, 035003 (2004)
6. M. Rodriguez, R. Bourayou, G. Méjean, J. Kasparian, J. Yu, E. Salmon, A. Scholz, B. Stecklum, J. Eislöffel, U. Laux, A.P. Hatzes, R. Sauerbrey, L. Wöste, J.-P. Wolf, Phys. Rev. E **69**, 036607 (2004)
7. S. Skupin, L. Bergé, U. Peschel, F. Lederer, G. Méjean, J. Yu, J. Kasparian, E. Salmon, J.P. Wolf, M. Rodriguez, L. Wöste, R. Bourayou, R. Sauerbrey, Phys. Rev. E **70**, 046602 (2004)
8. J.R. Penãno, P. Sprangle, B. Hafizi, A. Ting, D.F. Gordon, C.A. Kapetanakos, Phys. Plasmas **11**, 2865 (2004)
9. S. Champeaux, L. Bergé, Phys. Rev. E **71**, 046604 (2005)
10. S.L. Chin, S.A. Hosseini, W. Liu, Q. Luo, F. Théberge, N. Aközbek, A. Becker, V.P. Kandidov, O.G. Kosareva, H. Schroeder, Can. J. Phys. **83**, 863 (2005)
11. A. Ting, D.F. Gordon, E. Briscoe, J.R. Penãno, P. Sprangle, Appl. Opt. **44**, 1474 (2005)
12. A. Ting, I. Alexeev, D. Gordon, R. Fischer, D. Kaganovich, T. Jones, E. Briscoe, J. Penãno, R. Hubbard,P. Sprangle, Phys. Plasmas **12**, 056705 (2005)
13. G. Méchain, C.D. Amico, Y.-B. Andre, S. Tzortzakis, M. Franco, B. Prade, A. Mysyrowicz, A. Couairon, E. Salmon, R. Sauerbrey, Opt. Commun. **247**, 171–180 (2005)
14. R. Ackermann, G. Méjean, J. Kasparian, J. Yu, E. Salmon, J.-P. Wolf, Opt. Lett. **31**, 86 (2006)
15. F. Théberge, W. Liu, P.T. Simard, A. Becker, S.L. Chin, Phys. Rev. E **74**, 036406 (2006)
16. L. Bergé, S. Skupin, R. Nuter, J. Kasparian, J.-P. Wolf, Rep. Prog. Phys. **70**, 1633 (2007)
17. A. Couairona, A. Mysyrowicz, Phys. Rep. **441**, 47–189 (2007)

18. Y. Chen, F. Théberge, O. Kosareva, N. Panov, V.P. Kandidov, S.L. Chin, Opt. Lett. **32**, 3477 (2007)
19. S. Eisenmann, A. Pukhov, A. Zigler, Phys. Rev. Lett. **98**, 155002 (2007)
20. S. Champeaux, L. Bergé, Phys. Rev. E **77**, 036406 (2008)
21. Y. Ma, X. Lu, T.-T. Xi, Q.-H. Gong, J. Zhang, Opt. Express **16**, 8332 (2008)
22. J. Bernhardt, W. Liu, F. Théberge, H.L. Xu, J.F. Daigle, M. Châteauneuf, J. Dubois, S.L. Chin, Opt. Commun. **281**, 1268–1274 (2008)
23. J. Kasparian, J.-P. Wolf, Opt. Express **16**, 466 (2008)
24. L.B. Loeb, J.M. Meek, *The Mechanism of the Electric Spark* (Oxford University Press, Oxford, 1941)
25. Y.P. Raizer, *Gas Discharge Physics* (Springer, Berlin, 1991)
26. B.L. Fontaine, D. Comtois, C.-Y. Chien, A. Desparois, F. Génin, G. Jarry, T. Johnston, J.-C. Kieffer, F. Martin, R. Mawassi, H. Pépin, F.A.M. Rizk, F. Vidal, C. Potvin, P. Couture, H.P. Mercure, J. Appl. Phys. **88**, 610 (2000)
27. H. Peppin, D. Comtois, F. Vidal, C.Y. Chien, A. Desparois, T.W. Johnston, J.C. Kieffer, B. La Fontaine, F. Martin, F.A.M. Rizk, C. Potvin, P. Couture, H.P. Mercure, A. Boudiou-Clergerie, P. Lalande, I. Gallimberti, Phys. Plasmas **8**, 2532–2539 (2001)
28. S. Tzortzakis, B. Prade, M. Franco, A. Mysyrowicz, S. Huller, P. Mora, Phys. Rev. E **64**, 57401 (2001)
29. M. Rodriguez, R. Sauerbrey, H. Wille, L. Wöste, T. Fujii, Y.-B. André, A. Mysyrowicz, L. Klingbeil, K. Rethmeier, W. Kalkner, J. Kasparian, E. Salmon, J. Yu, J.-P. Wolf, Opt. Lett. **27**, 772 (2002)
30. D.F. Gordon, A. Ting, R.F. Hubbard, E. Briscoe, C. Manka, S.P. Slinker, A.P. Baronavski, H.D. Ladouceur, P.W. Grounds, P.G. Girardi, Phys. Plasmas **10**, 4530 (2003)
31. J. Kasparian, M. Rodriguez, G. Méjean, J. Yu, E. Salmon, H. Wille, R. Bourayou, S. Frey, Y.-B. André, A. Mysyrowicz, R. Sauerbrey, J.-P. Wolf, L. Wöste, Science **301**, 61 (2003)
32. T. Fujii, N. Goto, M. Miki, T. Nayuki, T. Shindo, K. Nemoto, IEEJ Trans. FM **125**, 765 (2005). (in Japanese)
33. N. Goto, M. Miki, T. Fujii, T. Nayuki, T. Sekiya, T. Shindo, K. Nemoto, IEEJ Trans. FM **125**, 1059 (2005). (in Japanese)
34. G. Méjean, R. Ackermann, J. Kasparian, E. Salmon, J. Yu, J.-P. Wolf, K. Rethmeier, W. Kalkner, P. Rohwetter, K. Stelmaszczyk, L. Wöste, Appl. Phys. Lett. **88**, 021101 (2006)
35. R. Ackermann, G. Mechain, G. Mejean, R. Bourayou, M. Rodriguez, K. Stelmaszczyk, J. Kasparian, J. Yu, E. Salmon, S. Tzortzakis, Y.-B. Andre, J.-F. Bourrillon, L. Tamin, J.-P. Cascelli, C. Campo, C. Davoise, A. Mysyrowicz, R. Sauerbrey, L. Woeste, J.-P. Wolf, Appl. Phys. B **82**, 561–566 (2006)
36. T. Fujii, M. Miki, N. Goto, A. Zhidkov, T. Fukuchi, Y. Oishi, K. Nemoto, Phys. Plasmas **15**, 013107 (2008)
37. J. Kasparian, R. Ackermann, Y.-B. André, G. Méchain, G. Méjean, B. Prade, P. Rohwetter, E. Salmon, K. Stelmaszczyk, J. Yu, A. Mysyrowicz, R. Sauerbrey, L. Wöste, J.-P. Wolf, Opt. Express **16**, 5757–5763 (2008)
38. T. Fujii, A. Zhidkov, M. Miki, K. Sugiyama, N. Goto, S. Eto, Y. Oishi, E. Hotta, K. Nemoto, Dynamics and kinetics of laser-filament plasma in strong external electric fields and applications. Chin. J. Phys. **52**, 440–464 (2014)
39. A. Houard, Y. Liu, B. Prade, V.T. Tikhonchuk, A. Mysyrowicz, Phys. Rev. Lett. **100**, 255006 (2008)
40. K. Sugiyama, T. Fujii, M. Miki, M. Yamaguchi, A. Zhidkov, E. Hotta, K. Nemoto, Opt. Lett. **34**, 2964 (2009)
41. K. Sugiyama, T. Fujii, M. Miki, A. Zhidkov, M. Yamaguchi, E. Hotta, K. Nemoto, Phys. Plasmas **17**, 043108 (2010)
42. T. Fujii, K. Sugiyama, M. Miki, A. Zhidkov, Y. Oishi, K. Nemoto, CRIEPI report H09020 (2010). (in Japanese)
43. T. Fujii, K. Sugiyama, M. Miki, A. Zhidkov, E. Hotta, K. Nemoto, J. Plasma Fusion Res. **86**, 669–677 (2010). (in Japanese)

9 Interaction of Femtosecond-Laser-Induced Filament Plasma

44. T. Fujii, in *Industrial Applications of Laser Remote Sensing*, ed. by T. Fukuchi, T. Shiina, Ch. 10 (Bentham Science Publishers, Sharjah, 2011)
45. E.P. Krider, Benjamin Franklin and lightning rods. Phys. Today **59**, 42–48 (2006)
46. C. Barthe, J.P. Pinty, Simulation of a supercellular storm using a three-dimensional mesoscale model with an explicit lightning flash scheme. J. Geophys. Res. **112**, D06210 (2007). doi:10.1029/2006JD007484
47. V.A. Rakov, M.A. Uman, *Lightning: Physics and Effects*, Ch. 7 (Cambridge Univ. Press, Cambridge, 2003)
48. S. Uchida, Y. Shimada, H. Yasuda, S. Motokoshi, C. Yamanaka, T. Yamanaka, Z. Kawasaki, K. Tsubakimoto, J. Opt. Technol. **66**, 199–202 (1999)
49. D.R. MacGorman, W.D. Rust, *The Electrical Nature of Storms*, Ch. 6 (Oxford Univ. Press, London, 2006)
50. T. Fujii, T. Fukuchi (eds.), *Laser Remote Sensing* (CRC Press, Boca Raton, 2005)
51. S.H. Lin, A.A. Villaeys, Y. Fujimura, *Advances in Multi-photon Processed and Spectroscopy*, vol. 16, Ch. 3 (World Scientific, Singapore, 2004)
52. J.-F. Daigle, G. Méjean, W. Liu, F. Théberge, H.L. Xu, Y. Kamali, J. Bernhardt, A. Azarm, Q. Sun, P. Mathieu, G. Roy, J.-R. Simard, S.L. Chin, Appl. Phys. B **87**, 749–754 (2007)
53. F.J. Mehr, M.A. Biondi, Phys. Rev. **181**, 264 (1969)
54. A. Zhidkov, A. Sasaki, Phys. Rev. E **59**, 7085–7095 (1999)
55. R. Salamé, N. Lascoux, E. Salmon, R. Ackermann, J. Kasparian, J.-P. Wolf, Appl. Phys. Lett. **91**, 171106 (2007)
56. G. Méjean, J. Kasparian, J. Yu, E. Salmon, S. Frey, J.-P. Wolf, S. Skupin, A. Vincotte, R. Nuter, S. Champeaux, L. Berge, Phys. Rev. E **72**, 026611 (2005)
57. G. Méchain, G. Méjean, R. Ackermann, P. Rohwetter, Y.-B. Andre, J. Kasparian, B. Prade, K. Stelmaszczyk, J. Yu, E. Salmon, W. Winn, L.A. (Vern) Schlie, A. Mysyrowicz, R. Sauerbrey, L. Wöste, J.-P. Wolf, Appl. Phys. B **80**, 785–789 (2005)

Chapter 10
Development of an Apparatus for Characterization of Cluster-Gas Targets for Laser-Driven Particle Accelerations

Satoshi Jinno, Yuji Fukuda, Hironao Sakaki, Akifumi Yogo, Masato Kanasaki, Kiminori Kondo, Anatoly Ya. Faenov, Igor Yu. Skobelev, Tatiana A. Pikuz, Alexy S. Boldarev and Vladimir A. Gasilov

---

S. Jinno (✉)
Nuclear Professional School, The Graduate School of Engineering, The University of Tokyo, 2-22 Shirakata-Shirane, Ibaraki, Kyoto 319-1188, Japan
e-mail: jinno@nuclear.jp

Y. Fukuda · H. Sakaki · A. Yogo · M. Kanasaki · K. Kondo · A. Y. Faenov · T. A. Pikuz
Kansai Photon Science Institute, Japan Atomic Energy Agency, 8-1-7 Umemidai, Kizugawa-city, Kyoto 619-0215, Japan
e-mail: jinno.satoshi@jaea.go.jp

Y. Fukuda
e-mail: fukuda.yuji@jaea.go.jp

H. Sakaki
e-mail: sakaki.hironao@jaea.go.jp

A. Yogo
e-mail: yogo.akifumi@jaea.go.jp

K. Kondo
e-mail: kondo.kiminori@jaea.go.jp

M. Kanasaki
Graduate School of Maritime Sciences, Kobe University, Kobe 658-0022, Japan
e-mail: kanasaki.masato@jaea.go.jp

A. Y. Faenov · I. Y. Skobelev · T. A. Pikuz
Joint Institute for High Temperatures, Russian Academy of Sciences, Izhorskaya 13 bld.2, Moscow 125412, Russia
e-mail: anatolyf@hotmail.com

I. Y. Skobelev
e-mail: igor.skobelev@gmail.com

T. A. Pikuz
e-mail: pikuz.tatiana@jaea.go.jp

A. S. Boldarev · V. A. Gasilov
Keldysh Institute of Applied Mathematics, Russian Academy of Science, Moscow 125047, Russia
e-mail: boldar@imamod.ru

K. Yamanouchi et al. (eds.), *Progress in Ultrafast Intense Laser Science XI*, Springer Series in Chemical Physics 109, DOI: 10.1007/978-3-319-06731-5_10, © Springer International Publishing Switzerland 2015

Abstract CO_2 clusters formed in supersonic expansion of a mixed-gas of CO_2/H_2 or CO_2/He through a three-staged conical nozzle have been verified by measuring the angular distribution of the light scattered from cluster target. The angular distribution is fitted by convolving a lognormal size distribution with the scattering coefficients calculated based on the Mie theory. The reliability of the size measurement is verified to be within an experimental error of 10 % using standard particles. The mean sizes of CO_2 clusters at the target center for the cases of CO_2/H_2 and CO_2/He gas mixtures are estimated to be 0.26 and 0.22 μm, respectively. For the CO_2/H_2 mixed-gas target, the variation of the mean cluster size inside the gas jet is constant within the experimental error. Furthermore, the cluster density is estimated to be 5.5×10^8 clusters/cm^2 by measuring the attenuation of the laser beam intensity. In addition, total gas density profiles in radial direction are obtained via the Abel inversion from the phase shift of the light passing through the target by utilizing an interferometer. The variation of the cluster mass fraction along the radial direction of the target is almost constant, which is consistent with a Boldarev's model.

10.1 Introduction

Realization of an advanced compact accelerator based on laser-driven ion acceleration attracts our interests during the last several years [1, 2] due to promising medical and industrial applications, e.g., cancer therapy [3–5], isotope preparation for medical applications [6], proton radiography [7], and controlled thermonuclear fusion [8]. We have focused attention on a cluster-gas target, which consist of clusters embedded in a background gas, in order to satisfy both the laser propagation into the plasma and existence of the near critical density plasma. In the laser-driven ion acceleration by using a cluster-gas targets, efficient generation of high energy ions up to 50 MeV per nucleon was achieved [9, 10] beyond traditional methods. In such experiments, the detailed knowledge about three independent parameters characterizing the cluster-gas target, i.e., cluster size, gas density, and cluster mass fraction, are necessary for understanding of acceleration mechanisms involved in the laser-cluster interactions and improvement of the acceleration efficiency.

Traditionally, a semi-empirical scaling parameter Γ^* proposed by Hagena [11] has been widely used to predict the mean cluster size since the first experimental demonstration of MeV ion generation [12]. In the framework of the Hagena's theory, the dimensionless parameter Γ^* completely defines all the parameters of the cluster targets in an averaged manner. The cluster-gas targets including several nanometer-sized clusters have been typically characterized by measuring the intensity of Rayleigh scattering of the light on clusters [13–16], yielding only a very approximate estimation. The Hagena's theory does not, however, take into account the thermodynamic properties of the background gas, which in fact plays a crucial role in laser-cluster interactions, including the long channel formation due to self-

---

V. A. Gasilov
e-mail: vgasilov@yandex.ru

10 Development of an Apparatus for Characterization of Cluster-Gas Targets

focusing [17], which can lead to generations of well collimated relativistic electrons and magnetic vortex [18]. Therefore, more detailed model for the clusters-gas targets, including the effects of the background gas, is required for the deep understanding of the laser-cluster interactions.

In this context, Boldarev et al. [19] has built a two-dimensional hydrodynamic model to evaluate the cluster-gas target parameter, which consists of the continuous gas phase and the discrete condensed phase (cluster). In this model, cluster mass fraction $f_c = N_\# N_c / (N_m + N_\# N_c)$, where $N_\#$ is the mean cluster size, N_m and N_c denote the number density of monomers and clusters, respectively, has been firstly introduced. The parameter f_c has been recently evaluated by measuring femtosecond evolution of the refractive index by single-shot frequency domain holography (FDH), giving a new concept for recovering the cluster size distribution [20].

By using simulations based on the Boldarev's model, a three-staged nozzle has been designed for the purpose of producing a sufficient amount of submicron-sized clusters at room temperature [21], which has been employed for the laser-driven ion acceleration experiments at JAEA-KPSI [9, 10]. The mean size of clusters produced by this nozzle is shown to be much larger than that of the Hagena's formula. Concerning the reliability of the Boldarev's model, however, it has been verified by Rayleigh scattering method only for several tens of nanometer-sized clusters [22]. In the case that the particle size is of the order of the visible light wavelength, i.e., the realm of Mie scattering, the angular distribution measurement of scattered light is needed. For example, the size measurements using Mie scattering have been performed for the micron-sized aerosol generated from the liquid [23, 24]. The Mie scattering method recently has been applied to submicron-sized clusters produced based on the Boldarev's model [25].

In present article, we give a full detail of the development of an apparatus capable of measuring micron-sized particles using Mie scattering and total gas densities using an interferometer with the second-harmonics of Nd:YAG laser pulses. The characterization of cluster-gas target, produced in supersonic gas expansion of CO_2/H_2 mixed gas from the three-staged nozzle, has been conducted by using the apparatus. These experimental results are compared with the Boldarev's model.

10.2 Experiments and Analysis for Characterization of the Cluster-Gas Target

10.2.1 Angular Distribution of Scattered light

Figure 10.1 shows the experimental setup for the measurements of the particle sizes using Mie scattering and the total gas density using the Michelson interferometer with a roof prism. The second harmonic (532 nm) of Nd:YAG laser (Spectra Physics, Quanta-ray) is split into the lines for the size measurement and the density measurement of the cluster-gas target. The laser beam with an energy below 1 μJ at a long pulse mode (several tens μs) is focused to the center of the vacuum chamber with

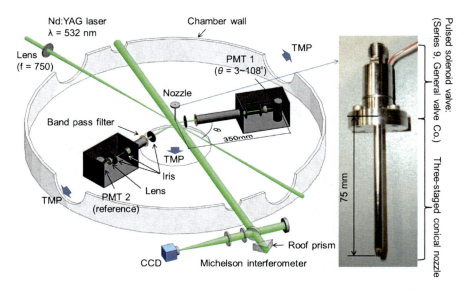

Fig. 10.1 Optical design for particle size measurements using Mie scattering and total gas density measurements using interferometry and photograph of a set of a pulsed solenoid valve and a three-staged conical nozzle

a plano-convex lens (f = 750 mm). The focal spot has 100 μm diameter at $1/e^2$ intensity and the Rayleigh length of 1 mm. The laser peak intensity is estimated much less than 1 kW/cm^2. Therefore, there is practically no probability for destruction of the clusters by the laser irradiation, considering CO_2 solid (dry ice) with the latent heat of vaporization of 370 kJ/kg and the absorption coefficient of several tens m^{-1} at the wavelength of 532 nm [26]. The polarization plane of the incident laser is perpendicular to the scattering plane. The laser pulse and gas injection timing are synchronized at 1 Hz, which is sufficiently slow rate to keep the vacuum condition which is evacuated by turbo molecular pumps (TMPs) backward-pumped with dry scroll pumps. A pulsed solenoid valve (Series 9, General valve Co.) controls the ejection of the gas, which becomes the supersonic jet through the three-staged conical nozzle. The opening time of the valve was set at 1 ms to ensure a stable gas flow. The pulsed solenoid valve and the nozzle are placed in room temperature.

In order to acquire angular distribution of the light scattered from the target, a photo-multiplier (PMT 1) is mounted on a movable stage from 3 to 108° at a distance of 350 mm from the target, where the 0° is defined as the laser propagation direction. Since the angular variation of Mie scattering becomes small with decreasing the particle size, which worsen the convergence of fitting process. If the measurements are truncated up to small angle, the average and the variance of the lognormal function, which determines the size distribution, are apt to be converged into local minimums. Therefore, our apparatus has a feature that the size distribution can be estimated accurately over a wide particle range by measuring the angular distribution three-times wider than others [23, 24]. Three irises on the line to the PMT 1 define

10 Development of an Apparatus for Characterization of Cluster-Gas Targets 219

the angular resolution of 0.5°. Another PMT 2 is placed perpendicular to the laser axis to detect a reference signal so as to compensate intensity fluctuations caused by the target samples. Energy fluctuations of incident laser pulses are monitored outside of the chamber. The scattered signal intensities are normalized shot by shot with the reference signal and the laser energy. The position dependence of the scattered signal intensity in this measurement can be obtained by shifting the nozzle position for the laser axis.

10.2.2 Derivation of Cluster Size Distribution

The size distribution of the clusters is analyzed in a following procedure: the scattering coefficient $F(x, \theta)$ strongly depends on the size parameter $x = 2\pi r n/\lambda$ according to the Mie scattering theory, where θ, r, n, and λ are scattering angle, particle radius, refractive index of medium ($=1$ in vacuum), and wavelength, respectively. The $F(x, \theta)$ is calculated applying an open Mie scattering code [27]. The size distribution of particle $N(x)$ is obtained by least-square fit of the angular distribution of the scattered light intensity $I(\theta)$ from following equation,

$$\frac{I(\theta)}{I_0} = \alpha \sum_x F(x, \theta)N(x), \tag{10.1}$$

where, α is the unknown apparatus constant, I_0 is the incident laser energy. In present study, $N(x)$ is assumed as a lognormal function [28], whose average and variance are obtained by fitting algorithm of the Nelder-Mead method. In present case, the $N(x)$ has the dimension of the surface density projected along the laser axis.

In the above procedure, only relative cluster size distribution shape $\alpha N(x)$ can be obtained. Then, the absolute cluster density $N(x)$ can be calculated by the following formula by measuring the transmitted light intensity I_t.

$$\frac{I_t}{I_0} = \exp\left(-\alpha^{-1} \sum_x C_{\text{ext}}(x)\alpha N(x)\right), \tag{10.2}$$

where, the $C_{\text{ext}}(x)$ is an extinction cross-section which can be calculated according to the Mie scattering theory. The α^{-1} is determined to satisfy this formula, and then the $N(x)$ can be found.

10.2.3 Spatial Distribution of Clusters

The nozzle can be scanned perpendicular to the laser axis in order to obtain the spatial distribution of clusters. As shown in Fig. 10.2, the angular distribution of the scattered light intensities $I(\theta, y_i)$ are measured at the position y_i on the axis perpendicular to the laser in the scattering plane. Because the PMT can accept the light scattered

Fig. 10.2 Schematic view of Abel inversion

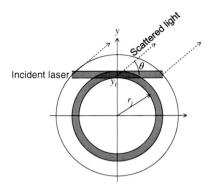

from the target in the overall laser pass, the scattering intensities $I(\theta, y_i)$ should be transformed into the emission coefficients $\varepsilon(\theta, r_j)$ in the radial direction r_j of the target by employing the simple discretization of Abel inversion as following equation [29],

$$\varepsilon(\theta, r_j) = -\frac{1}{\pi \Delta y} \sum_{i=j}^{N-1} \frac{I(\theta, y_{i+1}) - I(\theta, y_i)}{\sqrt{(i + \frac{1}{2})^2 + j^2}}, \tag{10.3}$$

where, Δy is the spatial separation between the data, which is the laser scan interval in this case. The radial separation Δr is expressed to be $\Delta r = \Delta y$. Using the $\varepsilon(\theta, r_j)$ instead of $I(\theta)$ of the (10.1), the number densities of clusters at the radial position, which has the dimension of the volume density, are obtained by the similar fitting procedure.

10.2.4 Total Gas Density Profile

For measuring the total gas density, the Michelson interferometer with a roof prism was adopted on another laser line with collimated beam as shown in Fig. 10.1. A probe light traversing the cluster-gas target medium interferes with a reference light passing the side of the target. The image is magnified by a telescope and recorded by a CCD camera. The interferogram produces a planar phase shift image resulting from the passage of the light across the cylindrically symmetric medium. The radial profile of the refractive index of the target $n(r_j)$ can be obtained via Abel inversion of this phase shift $\Delta\varphi(y_i)$ in the same manner as the (10.3), which is expressed as follows,

$$\frac{2\pi}{\lambda_0}\left(n(r_j) - 1\right) = -\frac{1}{\pi \Delta y} \sum_{i=j}^{N-1} \frac{\Delta\varphi(y_{i+1}) - \Delta\varphi(y_i)}{\sqrt{(i + \frac{1}{2})^2 + j^2}}, \tag{10.4}$$

10 Development of an Apparatus for Characterization of Cluster-Gas Targets 221

where, λ_0 is the wavelength of the incident laser light and the spatial separation Δy is determined by the spatial resolution of the acquired image in this case.

The refractive index is dominated by the density and the polarizability of the target particles. A cluster polarizability-volume α_c can be expressed as the product of the number of molecules per cluster $N_\#$ and molecular polarizability-volume α_m as follows [30, 31],

$$\alpha_c = N_\# \alpha_m. \tag{10.5}$$

In the present case, assuming the homogeneous mixed-gas of CO_2 and H_2, the relation between the radial refractive index and gas density can be modified as follows,

$$\frac{n^2 - 1}{4\pi} = N_c \alpha_c + (1 - \eta) N_m \alpha_m + N_H \alpha_H \tag{10.6}$$

$$= N_c N_\# \alpha_m + (1 - \eta) N_m \alpha_m + N_H \alpha_H \tag{10.7}$$

$$= N_m \alpha_m + N_H \alpha_H, \tag{10.8}$$

where, η is the fraction of molecules under cluster phase, N_c is the cluster density, N_m and N_H are the total molecular densities of CO_2 and H_2, and $\alpha_m = 2.59 \times 10^{-24}$ cm^3 and $\alpha_H = 0.802 \times 10^{-24}$ cm^3 are the polarizability-volumes averaged in terms of orientation of CO_2 and H_2, respectively. Also, total molecular densities have the following relationship to the partial pressures of mixed-gas,

$$N_m / N_m^0 : N_H / N_H^0 = p_m : p_H = 3 : 7, \tag{10.9}$$

where, N_m^0 and N_H^0 are the densities at the standard temperature and pressure, and p_m and p_H are the partial pressures of CO_2 and H_2, respectively. Finally, we can calculate the profile of the total molecular densities in radial direction from the (10.8) and (10.9) and using the refractive index derived in (10.4).

10.3 Results

10.3.1 Size Measurement of Standard Particles

The standard silica and the standard polystyrene particles (Thermo Fisher Scientific Inc.) suspended in water in a cylindrical quartz cell were used for the proof-of-principle experiments using Mie scattering. These catalog particle sizes are 0.49 (S.D. 0.02), 0.99 (S.D. 0.02), and 1.57 (S.D. 0.04) μm for the silica particles, 0.102 (S.D. 0.005), 0.240 (S.D. 0.004), 3.00 (S.D. 0.03) and 6.01 (S.D. 0.06) μm for the polystyrene particles, respectively. The 100 ppm by weight standard particles suspended in a cylindrical quartz cell filled with pure water having a 10 mm inner diameter are placed at the center of the chamber in air. Because the range of the target

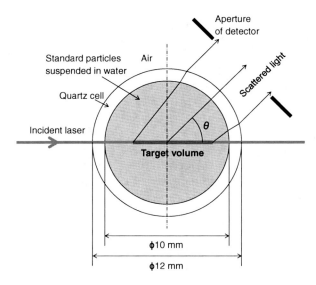

Fig. 10.3 Top view of the cylindrical quartz cell. It shows the diagram of correction of the target volume depending on the detection angle

cell is wider than the observational range of the detector as shown in Fig. 10.3, the intensities of scattered light were corrected for the attention target volume, which varies in accordance with detection angle including the effect of Snell's law.

Figure 10.4a–c shows the angular distributions of the light scattered from the silica and the polystyrene particles, respectively. The forward scattering ratio increases with the particle size, which is the characteristic of Mie scattering. The (10.1) gives fitting curves which reproduce well the angular distributions. As shown in Fig. 10.4a′–c′, the size distributions are in good agreement with the given sizes of the standard particles within an experimental error of 10 %, which ensures that our size measurement procedure well reproduces the original particle size.

The standard deviation is affected by the following things: the angular resolution of the detector which is 0.5°, the expansion of the incident laser in the medium due to refractions at the boundaries of the cylindrical quartz cell, and the stray light occurring at each boundary. These make the variation of scattering angular distribution gradual. The larger standard deviation of the size distribution is necessary to fit such experimental data. On the other hand, these had little effect on the mean size of particles.

10.3.2 Measurement of the Gas Jet Pressure

In the cases of the pulsed solenoid valve with and without the three-staged conical nozzle, the gas jet pressures were measured by tapping on an acceleration sensor by the gas jet. As shown in Fig. 10.5, the signals from the acceleration sensor were

10 Development of an Apparatus for Characterization of Cluster-Gas Targets   223

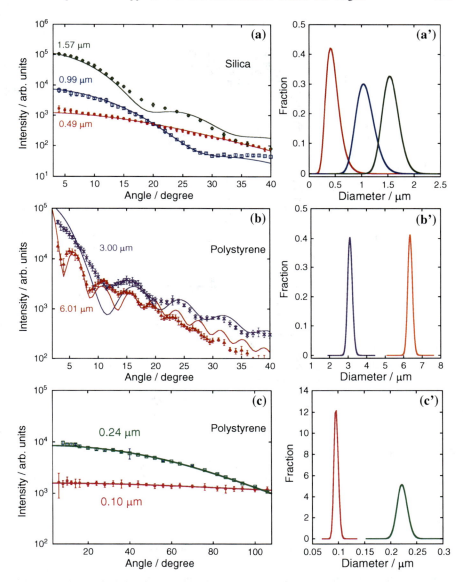

Fig. 10.4 The angular distribution of light scattered from the standard silica particles (**a**) and the standard polystyrene particles (**b**), (**c**). The lines are fitting curves convoled with the lognormal size distribution. The size distributions of the standard particles (**a'**), (**b'**), (**c'**) are estimated by the fitting procedure based on Mie theory, respectively

measured for H_2 (70%) + CO_2 (30%) mixed-gas at the stagnation pressures from 1 to 6 MPa, respectively. The signals for the worn-out (old) pulsed solenoid valve at 6 MPa were also measured for comparison. There is no variation between 3 and 6 MPa in

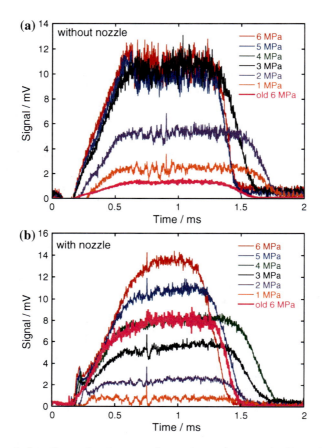

Fig. 10.5 Signals from the acceleration sensor by tapping on the sensor by the gas jet in the cases of the pulsed solenoid valve without (**a**) and with (**b**) the three-staged conical nozzle. The signals were measured for H_2 (70 %) + CO_2 (30 %) mixed-gas at the stagnation pressure from 1 to 6 MPa, respectively. The signals for the worn-out pulsed solenoid valve at 6 MPa were also measured for comparison

the case of only the pulsed solenoid valve. This result suggests that the gas flow angle spreads with increasing in the stagnation pressure. As proof of this, the pressure of the gas jet collimated through the nozzle has a linearity for the stagnation pressure. The low signal for the old valve without the nozzle is thought to be due to the gas flow with the wide angle. The signal of the gas jet with the nozzle is equivalent to that at 4 MPa in the case of the another valve. Thus, the each pulsed solenoid valve has individual character in regard to the spread angle and the pressure of the gas jet. We also found that these parameters depend on the degradation of a plastic poppet inside the pulsed solenoid valve and the stress to the plastic poppet for closing the valve. Therefore, the cluster formation may be affected by the valve lot and the adjustment in assembling the valve.

Fig. 10.6 a The angular distributions of scattered light from the CO$_2$ clusters at the radial position of 0, 0.2, 0.4, 0.6, 0.8, and 1.0 mm from the target center at 1 mm distance from the nozzle exit, respectively. **b** The lognormal size distributions estimated from the angular distributions of the scattered light

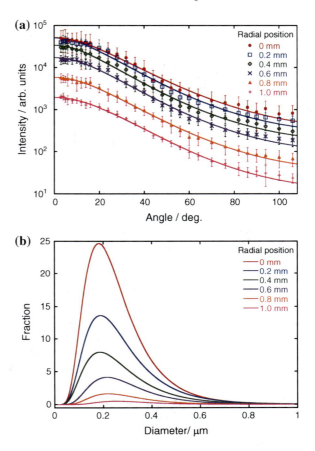

10.3.3 Characterization of H$_2$ (70 %) + CO$_2$ (30 %) Mixed-Gas Target

We have characterized the CO$_2$ clusters formed in the supersonic expansion of the H$_2$ (70 %) and CO$_2$ (30 %) mixed-gas through the three-staged conical nozzle. We have acquired the angular distributions of the scattered light at the positions of $y = 0, 0.2, 0.4, 0.6, 0.8$, and 1.0 mm from the target center by shifting the nozzle perpendicular to the laser axis at the distance of 1 mm from the nozzle orifice, respectively. As shown in Fig. 10.6a, the angular distributions at the positions divided radially were obtained through the Abel inversion of those raw data according to the (10.3). The distributions are featured by the forward scattering caused by Mie scattering, indicating that CO$_2$ clusters with the sizes of the laser wavelength scale exist in the cluster-gas target. Note that the contribution of Rayleigh scattering from monomer gases is usually negligible small because the scattering intensity is proportional to the sixth power of the particle size. As shown in Fig 10.6b, the lognormal size

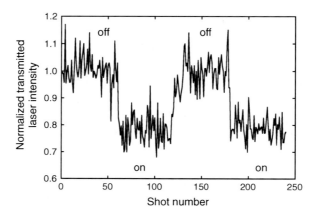

Fig. 10.7 Normalized transmitted laser intensity taken with no cluster-gas target present and with the cluster-gas target source on

distributions are obtained by the fitting procedure. The mean size of CO_2 clusters at the target center is estimated to be 0.26 μm. It is found that the variation of the mean cluster size inside the gas jet is constant within the experimental error. On the other hand, the Boldarev's model predicts that the mean cluster size is 0.588 μm with the standard deviation of 0.090 μm independent of the radial positions. The reason for this discrepancy will be discussed in Sect. 10.4.1. We note that the conventional Rayleigh scattering method could not be applicable for the estimation of our submicron clusters because the light scattered from such large clusters doesn't follow the simple sixth power law of the particle size, and has the angular distribution of forward scattering.

10.3.4 Cluster Density for H_2 (70 %) + CO_2 (30 %)

We can obtain the surface density of the clusters projected along the laser axis by measuring the transmission of the laser in the target as described in Sect. 10.2.2. Figure 10.7 shows the normalized transmitted laser intensity taken with no cluster-gas target present and with the cluster-gas target source on. Attenuation of the laser beam due to the presence of the cluster-gas target is clear. The mean transmission is found to be 0.787 ± 0.008 with the target, which results that the CO_2 cluster density is estimated to be 5.5×10^8 clusters/cm^2 or the CO_2 density of the cluster phase is corresponding to 1.1×10^{17} molecules/cm^2. Because the relative cluster concentration in the radial direction per unit length is found from the previous section, the surface density can be modified to the volume density of the clusters as a function of the radial direction as shown in Fig. 10.8.

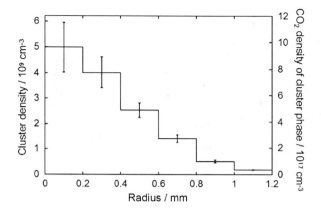

Fig. 10.8 Cluster density and CO_2 density of the cluster phase as a function of the radius from the target center axis at the distance $z = 1$ mm from the nozzle orifice

10.3.5 Total Gas Density Profile

Figure 10.9a shows the planar phase shift image near the nozzle exit using the Michelson interferometer. The density profile of the total molecules with cylindrical symmetry is obtained via Abel inversion of the phase shift image as shown in Fig 10.9b. Where, the radius is the distances from the center axis of the nozzle and the z is the distance from the nozzle orifice. As can be seen from Fig 10.9a and b, the gases diffuse in the radial direction with an increase in the distance of the z axis. As shown in Fig. 10.9c, the maximum densities of CO_2 and H_2 are 1.3×10^{19} and 2.9×10^{19} cm^{-3} at $z = 1.0$ mm, respectively. By contrast, the Boldarev's model predicts an approximately collimated gas flow at 7.5×10^{18} cm^{-3} and 1.7×10^{19} cm^{-3} for CO_2 and H_2 at $z = 1.0$ mm, respectively. Although the gas diffusion profiles obtained from the experiment and the model are different, the densities are consistent within a factor of two. The reason for the different of the gas density profile will be discussed in Sect. 10.4.1.

10.3.6 Cluster Mass Fraction

The cluster mass fraction f_c, i.e., the ratio of the CO_2 density of the cluster phase to the total density of CO_2 molecules, is expressed by the following equation,

$$f_c = N_\# N_c / (N_m + N_\# N_c). \tag{10.10}$$

where, the numerator of the right hand means the CO_2 density of the cluster phase in Fig. 10.8, and the denominator of the right hand also means the total density of CO_2

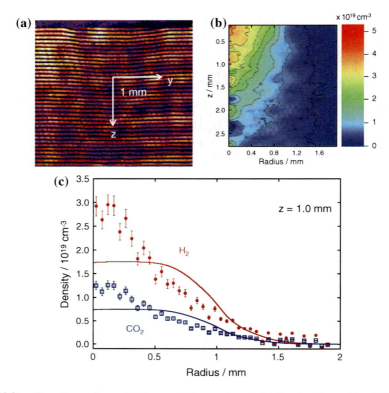

Fig. 10.9 **a** The planar phase shift image at the nozzle exit. **b** The density profile of the total molecules. z is the distance from the nozzle exit. The radius means the distance from the center axis of the nozzle. **c** The number densities of CO_2 and H_2 molecules at $z = 1$ mm. The lines indicate the results of the Boldarev's model

molecules in Fig. 10.9c. Figure 10.10 shows the variation of the cluster mass fraction for the radius from the center axis of the nozzle at $z = 1$ mm. We found that the cluster mass fraction is almost homogeneous regardless of the position of the gas expansion. Although the result of Boldarev's model lies greater than the experimental results, the tendency independent of the radius is consistent. The reason for this discrepancy will be discussed in Sect. 10.4.1.

10.3.7 Size Measurement of CO_2 Clusters in Helium (90 %) + CO_2 (10 %) Mixed-Gas Target

We have evaluated the size of CO_2 clusters formed in the supersonic expansion of 6 MPa mixed-gas of He (90 %) + CO_2 (10 %). The laser irradiation position is the target center ($y = 0$) at the distance of 1 mm from the nozzle orifice. Figure 10.11

10 Development of an Apparatus for Characterization of Cluster-Gas Targets

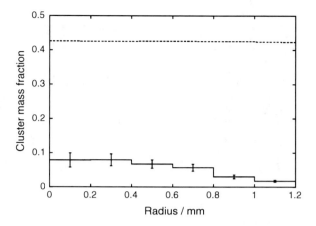

Fig. 10.10 The variation of the cluster mass fraction for the radius from the center axis of the nozzle at z = 1 mm. The broken line is the values by the Boldarev's model

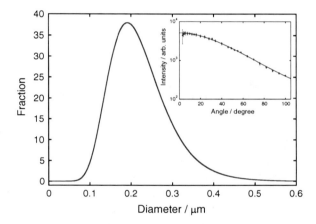

Fig. 10.11 The angular distributions of the light scattered from the CO_2 clusters in He (90 %) + CO_2 (10 %) gas targets for the stagnation pressure of 6 MPa (*top-right*). The lognormal size distribution estimated from the angular distribution

shows the angular distributions of light intensity scattered from CO_2 clusters and the cluster size distribution. The mean diameter of CO_2 cluster present in the laser axis is estimated to be 0.22 μm (S.D. 0.07 μm) by the fitting procedure. The Boldarev's model predicts that the size of CO_2 clusters is 0.360 μm (S.D. 0.072 μm) for the cases of CO_2/He. The difference between the experimental result and the model will be discussed in Sect. 10.4.1.

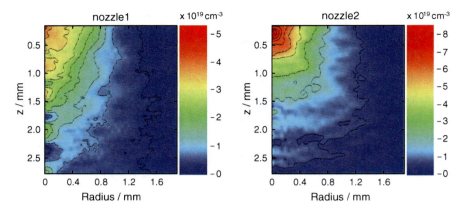

Fig. 10.12 Density profiles for two nozzles of same design

10.4 Discussion

10.4.1 Comparison with the Boldarev's Model

The discrepancies between experiment and simulation are recognized in the cluster size and its concentration. Its possible reasons are explained in this section. Firstly, the size measurements of the standard particles show that the total error coming from the measurement and the fitting processes is 10 %. Therefore, we think that the possible errors which originate from the measurements and the data fitting processes are small and cannot be the main reason for the discrepancy.

Secondly, concerning the uncertainties arising from the mathematical model of the cluster formation, it should be noted that the model, of course, has some its own limitations and assumptions. For example, it applies the thermodynamical approach to very small clusters of the liquid phase. There are some problems with adequate thermodynamical description of the metastable gas state (the clusters are formed in the supercooled gas). Therefore, we think that the inaccuracy of about 2 times may be not too high for such a model.

On the other hand, the reasons may be due to some geometrical inaccuracies in the nozzle shape. Different density profiles were observed for two nozzles of same design as shown in Fig. 10.12. Thus, the gas expansion conditions are very sensitive to the fabrication accuracy of the nozzle. For instance, we have performed the simulation by changing one of three inner diameters by only ± 100 μm, respectively. As a result, we find that the cluster sizes change from -30 to $+5$ %. If the parameters of the two or three inner diameters change at the same time, such change of the cluster size could appear more significantly. Accordingly, tiny variations in machining processes should be considered as a main reason for diminishing of the cluster size. In addition, the model set a boundary condition that is zero normal velocity to the nozzle's wall. When there is the roughness of the nozzle inside, the ideal gas flow assumed in the

10 Development of an Apparatus for Characterization of Cluster-Gas Targets

model is not realized. The degradation of the plastic poppet inside the pulsed solenoid valve may also change the ideal gas flow as described in Sect. 10.3.2. The possibility of these effects cannot be denied and may result in considerable diminishing of the cluster size.

Concerning the discrepancy of the cluster mass fraction between the model (40 %) and the experiment (10 %), the numerical simulations suggest that the cluster mass fraction changes from 41 to 49 % if one of three inner diameters varies by only ± 100 μm. Therefore, the discrepancy cannot be explained simply by tiny variations in the inner diameters. However, an evolution of turbulence flow inside the nozzle due a conical surface roughness could destroy clusters, which lowers the cluster mass fraction. This point will be discussed in future publications. Note that Gao et al. have evaluated the cluster mass fraction of 40 % by using their own gas jet nozzle, not by our three-staged conical nozzle, with the FDH technique [20].

As mentioned above, the assumption of the initial parameters in the model, the limitation of the thermodynamic approach, and the some inaccuracies in the nozzle fabrication are the main factors of the discrepancy between the measurements and the model. Since we have excluded the avoidable elements to a maximum extent, which could deteriorate experimental results, we believe that our procedure is enough reliable to compare experimental results to model. We have to always keep these points in mind in performing the experiments using the cluster-gas targets.

10.5 Summary

The apparatus for characterizing of the cluster-gas targets and the analytical methods by employing the Mie scattering method were developed. The proof-of-principle experiments were performed using the standard silica particles (0.49, 0.99, 1.57 μm) and the polystyrene particles (0.10, 0.24, 3.00, 6.01 μm). The reliability of the size measurement is verified to be within the experimental error of 10%, which ensures that our size measurement procedure well reproduces the original particle size.

The mean sizes of CO_2 clusters at the target center for the cases of CO_2/H_2 and CO_2/He gas mixtures were estimated to be 0.26 μm and 0.22 μm, respectively. For the CO_2/H_2 mixed-gas target, the variation of the mean cluster size inside the gas jet is constant within the experimental error. The cluster density is estimated to be 5.5×10^8 clusters/cm^2 by measuring the attenuation of the laser beam intensity. The total gas densities were measured by the Mickelson type interferometer with the roof prism. It was found that the gas diffusion profile is affected by the fabrication accuracy of the nozzle. The gas densities are verified to agree with the Boldarev's model within a few factors. The cluster mass fractions were spatially derived in combination with the Mie scattering method, the attenuation measurement of the laser beam intensity, and the interferometry. The variation of the cluster mass fraction in radial direction of the target is almost constant. The constant distributions of the mean cluster size and the cluster mass fraction inside the gas jet support the Boldarev's model, which claims the homogeneous cluster formation. Therefore, the model is reliable enough for the production of submicron-sized clusters.

In laser-driven ion acceleration experiment with the cluster-gas target using the J-KAREN Ti:sapphire laser at JAEA-KPSI, we found that electron energy spectra and shadow graphs of the target plasma sensitively depend on the focal positions inside the cluster-gas target. We believe that the knowledge of the initial state of the spatial distribution of the cluster-gas target provides the valuable information for understanding the elementary process of the laser-cluster interactions. We will discuss the relationship among the initial target state, the electron acceleration, and the ion acceleration in the future chapter.

Acknowledgments This work was supported by the Funding Program for Next Generation World-Leading Researchers (NEXT Program) from the Japan Society for the Promotion of Science (JSPS) and the grant RFBR 12-02-91169-GFEN-a.

References

1. A. Macchi, M. Borghesi, M. Passonio, Rev. Mod. Phys. **85**, 751 (2013)
2. H. Daido, M. Nishiuchi, A.S. Pirozhkov, Rev. Prog. Phys. **75**, 056401 (2012)
3. S.V. Bulanov, V.S. Khoroshkov, Plasma Phys. Rep. **28**, 453 (2002)
4. T. Tajima, D. Habs, X. Yan, Accel. Sci. Tech. **2**, 201 (2009)
5. A. Yogo, K. Sato, M. Nishikino, M. Mori, T. Teshima, H. Numasaki, M. Murakami, Y. Demizu, S. Akagi, S. Nagayama, K. Ogura, A. Sagisaka, S. Orimo, M. Nishiuchi, A.S. Pirozhkov, M. Ikegami, M. Tampo, H. Sakaki, M. Suzuki, I. Daito, Y. Oishi, H. Sugiyama, H. Kiriyama, H. Okada, S. Kanazawa, S. Kondo, T. Shimomura, Y. Nakai, M. Tanoue, H. Sasao, D. Wakai, P.R. Bolton, H. Daido, Appl. Phys. Lett. **94**, 181502 (2009)
6. I. Spencer, K. Ledingham, R. Singhal, T. McCanny, P. McKenna, E. Clark, K. Krushelnick, M. Zepf, F. Beg, M. Tatarakis, A. Dangor, P. Norreys, R. Clarke, R. Allott, I. Ross, Nucl. Instrum. Meth. Phys. Res. B **183**, 449 (2001)
7. M. Borghesi, D.H. Campbell, A. Schiavi, M.G. Haines, O. Willi, A.J. MacKinnon, P. Patel, L.A. Gizzi, M. Galimberti, R.J. Clarke, F. Pegoraro, H. Ruhl, S. Bulanov, Phys. Plasmas **9**, 2214 (2002)
8. M. Roth, T.E. Cowan, M.H. Key, S.P. Hatchett, C. Brown, W. Fountain, J. Johnson, D.M. Pennington, R.A. Snavely, S.C. Wilks, K. Yasuike, H. Ruhl, F. Pegoraro, S.V. Bulanov, E.M. Campbell, M.D. Perry, H. Powell, Phys. Rev. Lett. **86**, 436 (2001)
9. Y. Fukuda, A.Y. Faenov, M. Tampo, T.A. Pikuz, T. Nakamura, M. Kando, Y. Hayashi, A. Yogo, H. Sakaki, T. Kameshima, A.S. Pirozhkov, K. Ogura, M. Mori, T.Z. Esirkepov, J. Koga, A.S. Boldarev, V.A. Gasilov, A.I. Magunov, T. Yamauchi, R. Kodama, P.R. Bolton, Y. Kato, T. Tajima, H. Daido, S.V. Bulanov, Phys. Rev. Lett. **103**, 165002 (2009)
10. Y. Fukuda, H. Sakaki, M. Kanasaki, A. Yogoa, S. Jinno, M. Tampo, A. Faenov, T. Pikuz, Y. Hayashi, M. Kando, A. Pirozhkov, T. Shimomura, H. Kiriyama, S. Kurashima, T. Kamiya, K. Oda, T. Yamauchi, K. Kondo, S. Bulanov, Radiat. Meas. **50**, 92 (2013)
11. O.F. Hagena, Rev. Sci. Instrum. **63**, 2374 (1992)
12. T. Ditmire, J.W.G. Tisch, E. Springate, M.B. Mason, N. Hay, R.A. Smith, J. Marangos, M.H.R. Hutchinson, Nature **386**, 54 (1997)
13. T. Ditmire, E. Springate, J.W.G. Tisch, Y.L. Shao, M.B. Mason, N. Hay, J.P. Marangos, M.H.R. Hutchinson, Phys. Rev. A **57**, 369 (1998)
14. V. Kumarappan, M. Krishnamurthy, D. Mathur, L.C. Tribedi, Phys. Rev. A **63**, 023203 (2001)
15. K.Y. Kim, V. Kumarappan, H.M. Milchberg, Appl. Phys. Lett. **83**, 3210 (2003)

16. S. Sakabe, S. Shimizu, M. Hashida, F. Sato, T. Tsuyukushi, K. Nishihara, S. Okihara, T. Kagawa, Y. Izawa, K. Imasaki, T. Iida, Phys. Rev. A **69**, 023203 (2004)
17. I. Alexeev, T.M. Antonsen, K.Y. Kim, H.M. Milchberg, Phys. Rev. Lett. **90**, 103402 (2003)
18. T. Nakamura, S.V. Bulanov, T.Z. Esirkepov, M. Kando, Phys. Rev. Lett. **105**, 135002 (2010)
19. A.S. Boldarev, V.A. Gasilov, A.Y. Faenov, Tech. Phys. **49**, 388 (2004)
20. X. Gao, X. Wang, B. Shim, A.V. Arefiev, R. Korzekwa, M.C. Downer, Appl. Phys. Lett. **100**, 064101 (2012)
21. A.S. Boldarev, V.A. Gasilov, A.Y. Faenov, Y. Fukuda, K. Yamakawa, Rev. Sci. Instrum. **77**, 083112 (2006)
22. F. Dorchies, F. Blasco, T. Caillaud, J. Stevefelt, C. Stenz, A.S. Boldarev, V.A. Gasilov, Phys. Rev. A **68**, 023201 (2003)
23. A.P. Higginbotham, O. Semonin, S. Bruce, C. Chan, M. Maindi, T.D. Donnelly, M. Maurer, W. Bang, I. Churina, J. Osterholz, I. Kim, A.C. Bernstein, T. Ditmire, Rev. Sci. Instrum. **80**, 063503 (2009)
24. T.D. Donnelly, J. Hogan, A. Mugler, M. Schubmehl, N. Schommer, A.J. Bernoff, S. Dasnurkar, T. Ditmire, Rev. Sci. Instrum **76**, 113301 (2005)
25. S. Jinno, Y. Fukuda, H. Sakaki, A. Yogo, M. Kanasaki, K. Kondo, A.Y. Faenov, I.Y. Skebelv, T.A. Pikuz, A.S. Boldarev, V.A. Gasilov, Appl. Phys. Lett. **102**, 164103 (2013)
26. S.G. Warren, Appl. Opt. **25**, 2650 (1986)
27. W.J. Wiscombe, Mie scattering calculations: Advances in technique and fast, vector-speed computer codes. NCAR Technical Note 140 + STR (1979)
28. M. Kerker, *The Scattering of Light and Other Electromagnetic Radiation* (Academic Press, New York and London, 1969)
29. R. Álvarez, A. Rodero, M.C. Quintero, Spectrochim. Acta Part B **52**, 1665 (2002)
30. H.M. Milchberg, S.J. McNaught, E. Parra, Phys. Rev. E **64**, 056402 (2001)
31. H.-Y. Kim, J.O. Sofo, D. Velegol, M.W. Cole, G. Mukhopadhyay, Phys. Rev. A **72**, 053201 (2005)

Index

A

Abel inversion, 220
Above-threshold-ionization (ATI), 28, 30, 32, 34
    five-photon, 32, 34
Absorbing boundary, 107
Accuracy
    initial CEP, 190
Acetylene, 49
Acousto-optic programmable dispersive filters, 74
Alignment, 53
Alignment pulse, 54
Amplitude shaping, 75
Angular dependence of the ionization probability, 53
Asymmetric HHe^{2+}, 133
Asymmetry, 159, 162, 163, 166
Atomic units, 99

B

Boldarev's model, 217, 226–230
Bond-breaking, 1
Bond-making, 1
1,3-butadiene, 49

C

C–H stretch motion, 67
Carrier envelope phase (CEP), 45, 46, 149, 159, 162, 176
Channel closing, 35
Channel radius R_c, 99
CH_3^+ ions, 10, 18
Chirp, 2, 4, 8, 11
Circularly polarized attosecond pulses, 121

Cluster polarizability-volume, 221
Cluster size distribution, 219, 229
Cluster-gas target, 216, 217, 226
Coherent control, 73
Coherent electron wave packets (CEWPs), 149, 150, 159, 169
Coherent population control, 82
Coherent transient, 89
Coincidence analysis, 64
COLTRIMS, 47
Complete fragmentation, 61
Constructive quantum interference, 89
Control, 10, 11, 14, 15
Controlling molecular fragmentation reactions, 45
Corona, 198, 199, 201, 205
Correlation, 114
Coulomb explosion, 10, 64
Coulomb scattering, 104
Coupled electronic and nuclear dynamics, 68
Critical internuclear distance, 61, 67
Current density, 183, 184

D

Dication, 2
Direct path and sequential path, 83
Discharge
    corona, 204, 206
    streamer, 196
Discharge triggering, 196
Double ionization, 50
Dynamic Stark effect, 84

K. Yamanouchi et al. (eds.), *Progress in Ultrafast Intense Laser Science XI*,
Springer Series in Chemical Physics 109, DOI: 10.1007/978-3-319-06731-5,
© Springer International Publishing Switzerland 2015

Index

E
Electric
 field measurement, 195, 196
Electronically excited state surfaces, 51
Energetic protons, 65
Energy correlation, 36, 38
 maps, 36
Energy level diagram, 49
Enhanced ionization, 61
Entanglement, 114
Ethylene, 49
Excited state, 51
Extinction cross-section, 219
Extreme ultraviolet (XUV), 119, 121, 125, 143, 149, 150

F
Far field emission, 188
Femtosecond interferometry, 14
Few-cycle laser pulses, 46
Few-electron system, 101
Field measurement, 198, 211
Filament, 197, 201–205, 207, 208, 210
Filamentation, 197
Finite element, 107
Four-body coincidence analysis, 65
Fragmentation pathways, 45
Franck-Condon overlap, 29, 32
Freeman resonance, 32
Frequency de-tuning, 76
Frequency resolved optical grating (FROG), 3, 15, 179

G
Gas density profile, 220, 227
Genetic algorithm, 5, 15

H
Hagena's theory, 216
Half-revival, 57
H_3^+ formation, 15
High-intensity
 femtosecond laser, 197
High-order harmonic generation (MHOHG), 120, 150
H_3^+ ion, 1, 13
Hydrogen migration, 2

I
Inhomogeneous problem, 105

Initial CEP, 187
Interference, 145, 153, 155, 158, 162, 169
Interferogram, 220
Internal state distribution, 39
 $C_2H_5OH^+$, 39
Inverse overlap, 105
Inversion
 of THz waveform , 189
Ion-ion-coincidence, 2
Ions
 backward scattered, 7
 forward scattered, 7
Isomerization reactions, 53

J
J-KAREN, 232

K
Keldysh parameter, 34–36
Kerr self-focusing, 197, 202
Kinetic energy distribution, 17, 19, 36, 38
 CH_2OH^+, 36
 $C_2H_5^+$, 38
 Boltzmann-type distribution, 36
 fragment ions, 36
Kinetic energy release (KER) spectrum, 2, 59
Kramers-Henneberger, 98

L
Laser
 femtosecond laser, 195
 filament, 195–197, 199, 204–206
Laser field ionization, 197
Laser induced electron diffraction (LIED), 119–121, 129, 133, 143, 154
Laser pulse shaping, 74
Laser-driven ion acceleration, 216, 232
Laser-induced discharge, 201, 204
Lower lying orbital, 51

M
Magnus exponentiation, 107
MAPDs, 125
Method gain, 107
Michelson interferometer, 217, 220, 227
Mie scattering, 217–219, 221, 222, 225
Modified Gouy phase shift, 185
Molecular above threshold ionization (MATI), 119, 120, 139, 160
Molecular orbital, 124

Index

Molecular photoelectron angular distributions (MPADs), 119, 134, 143, 145, 151, 153, 159, 163
Molecular polarizability-volume, 221
Molecular tunneling theory, 56
Multi-bond version of EI, 68
Multi-bond version of the EI mechanism, 61
Multi-particle breakup, 104
Multiphoton, 197

N

Nitrogen (N_2) fluorescence, 204

O

One-photon de-excitation, 88
Optical parametric amplifier (OPA), 176
  three-stage, 177
Optical pulse shaper, 77
Orientation, 141

P

Particle-in-cell (PIC), 198
Perturbation, 162
Perturbation theory, 98
Perturbative, 166
Phase function programming, 92
Phase shaping, 75
Phase shift, 184
Phase space, 98
Photocurrent model, 175, 185
Photoelectron-photoion coincidence, 39
  coincidence condition, 26
  momentum imaging, 25, 39
Photoelectron-photoion coincidence (PEPICO), 24, 34
Plasma
  femto-second laser, 195
  filament, 195–200, 202, 203, 206, 210, 211
  measurement, 198
Plastic poppet, 224, 231
Polyatomic molecules, 44
Ponderomotive energy shift, 27, 35, 38
Ponderomotive potential, 97
Position-sensitive detectors
  delay-line anodes, 26
  microchannel plates, 26
Position-sensitive detectors (PSD), 25, 26
Propagation equation, 183, 187
Proton energy spectra, 63
Proton migration reaction, 56

Pulse shaping, 14
Pulse shaping methods, 74
Pulsed solenoid valve, 218, 222, 231
Pulses, 176

R

Rayleigh scattering, 216, 225
Re-collision, 95
Recollision ionization, 50
Recollision-induced excitation, 52
Region
  bound, 101, 102
  doubly ionized, 102, 104
  singly ionized, 101, 102
Remote
  field measurement, 195, 196
Remote measurement, 196, 198, 211
Rotation, 151
Rotational wavepacket, 54
Rydberg, 32
  3s, 32

S

Scaling
  curse of dimensions, 97
  in space, 97
  in time, 97
  infrared curse, 97
Scattering electron, 145, 155
Second-order perturbation calculation, 91
Sequential field ionization, 60
Simple man's model, 49
Single-ionization channel, 102
Spatial light modulator, 3, 74
Spectra
  $2 \times 1d$, 110
  double emission, 113
  elliptic polarization, 110
  long range, 109
  short range, 107
Spectral amplitude
  double, 105
  single, 99, 103
Spectral blocking, 78
Spectral linear chirp, 76
Spectral phase, 4
Spectral phase step, 76
Spectral quadratic chirp, 76
Stagnation pressure, 223
Standard particle, 221, 230
Stereo-ATI phase-meter, 47

S

Streamer, 198, 203, 209
Strong-field ionization, 44
Subsequent electronic excitation, 38
Supercontinuum, 179
Surprisingly high charge states, 61

T

Three-level ladder system, 82
Three-staged nozzle, 217
THz radiation, 180
    waveform-controlled, 176
Ti: sapphire laser, 200
Time dependent perturbation calculation, 81
Time-of-flight-mass spectrometer, 6
ToF distribution, 7
Transient photocurrent, 182, 192
Transition probability amplitude, 78
Tunnel ionization, 197

Tunneling theory, 52
Two-dimension Fourier transform spectroscopy, 81
Two-electron system, 96, 101

U

Ultra-fast quantum control, 73

V

Valence electrons, 44
Variation of CEP, 186, 189
Vector potential, 111
Velocity gauge, 98, 99
Vinylidene configuration, 56
Volkov solution, 98, 99
V-type transition, 81